Ulrich Schulz

Accelerated Testing

Nature and Artificial Weathering in the Coatings Industry

Ulrich Schulz: Accelerated Testing:
Nature and Artificial Weathering in the Coatings Industry

ISBN: 978-3-86630-908-1

Cover: Q-Lab Corp., Cleveland/USA

Ulrich Schulz
Accelerated Testing: Nature and Artificial Weathering in the Coatings Industry
Hannover: Vincentz Network, 2009
(European Coatings Fech Files)
ISBN 3-86630-908-2
ISBN 978-3-86630-908-1

Please ask for our book catalogue
Vincentz Network, Plathnerstr. 4c, 30175 Hannover, Germany
T (202) 684-6630, F (202) 380-9129
E-mail: books@american-coatings.com, www.american-coatings.com

Layout: Maxbauer & Maxbauer, Hannover, Germany

ISBN 3-86630-908-2
ISBN 978-3-86630-908-1

Ulrich Schulz

Accelerated Testing

Nature and Artificial Weathering in the Coatings Industry

Ulrich Schulz: Accelerated Testing:
Nature and Artificial Weathering in the Coatings Industry

ISBN: 978-3-86630-908-1

Foreword

"Accelerated Testing: Nature and Artificial Weathering in the Coatings Industry" is aimed at all those involved or interested in creating, producing, applying, and testing modern high-quality coatings for outdoor use.

Coatings are exposed to a great many severe natural stresses that cause a gradual deterioration of the properties which are responsible for the coatings' very quality. Nevertheless, buyers expect coated products to remain in an as-new condition - which is mostly characterised by a highly attractive appearance and intact surface - for as long as possible. This calls for coatings of high weatherability and long service life. In this book, accelerated testing, through its simulation of the destructive action of natural weathering, is the means for testing this coating quality. Test engineers shoulder much responsibility because not only must the results form the basis for reliable predictions, but they must also be obtained economically and as quickly as possible. Their results are the dominant factor in any decision to take a new coating creation into series production.

Accelerated testing has become an indispensable tool in the paint and coatings chemistry as a means of avoiding nasty surprises by coatings in normal use. Other methods of predicting service life are still too unreliable, given the extent of current weathering knowledge. Modern-day, high-quality coatings are highly complex systems which contain numerous essential additives. Not surprisingly, coatings chemistry is therefore sometimes jokingly likened to alchemy. But natural weathering, in all its random manifestations of different impact, is equally complex.

Words alone cannot describe how best to simulate the team-like interaction of such a complex system in the laboratory. There is more to successful simulation than applying a standardized test method, or switching on a fully controlled weathering device which has been marketed as an all-rounder. It takes know-how, experience and skill. This book will help such abilities to be acquired.

The culmination of decades of experience in the development and performance of accelerated testing methods, it will help those engaged in coatings testing to successfully introduce and implement accelerated testing. It will help them not only to critically assess and identify the weaknesses of accelerated testing methods but also to nimbly sidestep the likely errors and have confidence in the test results.

Berlin, Germany, August 2008

Ulrich Schulz

Ulrich Schulz: Accelerated Testing:
Nature and Artificial Weathering in the Coatings Industry

ISBN: 978-3-86630-908-1

Acknowledgements

With the onset of retirement, I at last found time to record the experiences I gained over many years in the specialised field of coatings weathering. First, I would like to thank the team of the former laboratory for "Climatic Resistance of Polymers" at the Federal Institute of Materials Testing and Research (BAM). Special thanks are due to Peter Trubiroha, former head of the laboratory, for our good teamwork. I am also indebted to Mr. Trubiroha's and my successors in this field at the BAM, Dr. A. Geburtig and Dr. V. Wachtendorf, for helping with the development of the book, especially by providing access to their latest research papers and discussing the latest advances in coatings testing. A. Sobota, M. Lorenz, and M. Schneider also served as a vast source of extensive experimental experience through their technical support over the years - thank you. Finally, many thanks to the members of the "Kurzbewitterung" working group at the VDA. As a permanent guest at its meetings, I was able to sit in on and fully participate in fruitful discussions between specialists representing paint makers and coatings users alike who deal directly with the problems of short-term weathering. These provided with me invaluable experience and inspiration for this book.

Ulrich Schulz: Accelerated Testing:
Nature and Artificial Weathering in the Coatings Industry

ISBN: 978-3-86630-908-1

Content

Ulrich Schulz: Accelerated Testing:
Nature and Artificial Weathering in the Coatings Industry

ISBN: 978-3-86630-908-1

1 Introduction

Ever since polymeric materials, to which organic coatings belong, have been used outdoors, there have been complaints about their poor weatherability. In order to gain knowledge about this important element of a coating's quality before it is actually used, it has become standard practice over the last 100 years to conduct weathering tests, either outdoors or in the laboratory. Performing such tests as efficiently as possible has become the aim and purpose of accelerated testing. More correctly, it serves to determine the coating's ageing behaviour. Deservedly, accelerated testing has evolved into the most important test method for finding out such important quality features of coatings as **weatherability** and **longevity**. Accelerated testing has become an indispensable tool of quality assurance for monitoring coating development as well as series production. Quality of a coating means giving the coated product the capability to maintain all the demanded surface properties for a specified or appropriate period of normal use. Metal surfaces of products made, for instance, of steel or aluminium, have to be protected against corrosion, wooden surfaces mostly against weather-induced disintegration and concrete surfaces predominantly against penetration of water, to name only a few examples. Besides this protective function, coatings have many others to perform. In many application areas, such as the building and automotive industries, coatings have to render the coated product decorative. A decorative appearance is one that is aesthetic and attractive. Coatings should be capable of positively influencing the buying habits of potential users. Besides this decorative function, special properties are required for certain areas of use. For instance, coatings for traffic signs and road markings have to be striking and clearly recognisable. Coatings for military use must provide the stipulated level of camouflage and be easy to decontaminate.

Longevity, often expressed by the term service life, describes how long the coating retains its necessary functionality or properties in real outdoor conditions, which are referred to in this book as normal use. This quality term is frequently characterised by a voluntary guarantee period. Guarantee periods of 10 years - demanded by users and provided by the paint makers - are common for state-of-the-art high-quality coatings. Such a long guarantee period is readily fulfilled by indoor coatings, e.g. furniture and floor coverings. However, it is still problematic for outdoor coatings, and remains a serious challenge for coatings developers. We need only think of coatings on building facades, metal bridges, cars, aircrafts and trains, all of which are attacked by various combinations of weathering stress (or weathering stress factors), and all of which must resist these influences for an appropriate period without suffering extensive damage! As almost all weathering stress has its origin in the natural environment, its action on the coating surface is summarized by the term weathering stress in this book. In simple terms, influences that are predominantly anthropogenic, such as acid rain and aggressive dust are also involved in weathering. How to prove that the coatings' quality is sufficient to resist all these weathering stresses? Accelerated testing is the most important method for testing this. It is expected to prove whether the coating to be tested is suitable for intended normal use. How well accelerated testing fulfils this task will be one of the main subjects of this book.

In this context, this book is intended to pass on know-how in the use and development of methods for accelerated testing. This know-how has been gained by a scientist who actively worked in this field for many years. Since special emphasis has been placed on drawing from practical experience and providing recommendations and tips for performing accelerated testing, the book has

Ulrich Schulz: Accelerated Testing:
Nature and Artificial Weathering in the Coatings Industry

ISBN: 978-3-86630-908-1

intentionally been kept short. However, the theoretical part is of sufficient scope to give the reader appropriate basic information for understanding coating ageing. The book is intended to give all persons involved in accelerated testing expertise in selecting the best test method and technique for the task at hand. It is also aimed at helping those staff to better recognise and avoid frequent mistakes that occur unnoticed or subconsciously in accelerated testing. As often reported, testing engineers have had the painful experience that mistakes during accelerated testing led to weathering results which were entirely useless for assessing the quality of the coatings to be tested. The book is also intended to sharpen the reader's ability to judge and critically assess any new weathering test methods in order that he may avoid useless work or wasted time, as well as false quality predictions that can lead to a loss of confidence in the testing laboratory.

Anyone involved in the development of coatings needs the tools to assess the strengths and weaknesses of accelerated testing so that the results can be used to predict the service life of the coatings to be tested. Nowadays, the uncertainty of the weathering results obtained is generally regarded to be indicative of the quality of the testing laboratory. Thus, this book is also intended to alert laboratory staff to the fact that its weathering results are crucial to the decision as to whether a new coating creation goes into series production or not.

2 General aspects of accelerated testing

2.1 Irradiation versus weathering

In the context of accelerated testing, the terms irradiation and weathering are frequently encountered. The requirements that must be fulfilled to label a test weathering can be properly derived from these two terms. First, let us look at Figure 2.1. It shows the essential weathering stress factors that act on outdoor coating surfaces, and illustrates the complexity of weathering stress. Due to the special properties of polymeric materials, to which organic coatings belong, solar radiation dominates the represented stress factors. This weathering stress factor is generally called global radiation.

More precisely, **global radiation** describes the totality of solar radiation, both direct and diffuse, that impinges on a horizontal surface. The direct radiation comes from the sun. The diffuse radiation is that portion which is scattered by the sky and clouds. In a departure from the precise definition, the term global radiation, as used in this book, is taken to mean the total natural radiation irrespective of the angle of incidence at the exposed surface. However, angles of incidence are quoted in readings for global radiation. Like global radiation, the oxygen contained in the ambient air can also be considered a predominant weathering stress factor. This explains why global radiation and oxygen are emphasised in Figure 2.1. As described in more detail in Chapter 3, the result of the combined action of global radiation and atmospheric oxygen, besides being expressed by the general term **coating ageing** is also referred to as **photooxidation**.

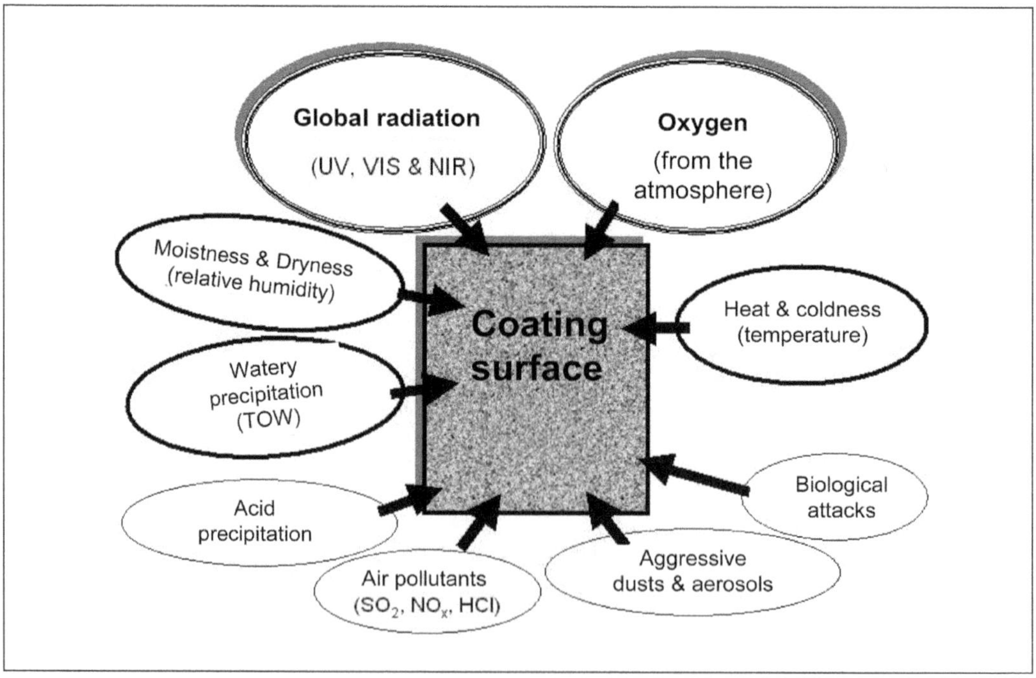

Figure 2.1: Weathering stress factors acting on a coating surface

Ulrich Schulz: Accelerated Testing:
Nature and Artificial Weathering in the Coatings Industry

ISBN: 978-3-86630-908-1

Figure 2.1 shows a number of other weathering stress factors that mostly act in combination with global radiation and atmospheric oxygen. Let us return to the terms weathering and irradiation. The difference between these two terms can be found in the answer to the question as to which and how many of the weathering stress factors listed in Figure 2.1 are included in the test for simulating environmental conditions. As regards accelerated testing, atmospheric oxygen can be considered a natural constant. The term irradiation is thus limited to the weathering stress factor irradiation, i.e. to only one of the weathering stresses. Therefore, a test method can be considered irradiation only if the irradiance is a test parameter which can be regulated. In this case, the test conditions are regulated either via the change in solar radiation or changes in artificial sources in irradiance. Temperature and relative humidity are a function of the irradiance acting on the coating surface and the surrounding space. Weathering clearly is derived from the term weather. Proper simulation of the action of the weather in normal use must involve other important weathering stresses, including the stress factors heat (as expressed by the ambient temperature), and water (by relative humidity and amount of rain) in addition to global radiation. Accelerated testing that means fulfilling these requirements. A good example of the difference between weathering and irradiation is the exposure of cars with locked doors and windows to the sun acting as radiation source. The car body coating is subjected to all outdoor weathering stresses acting in real combination. That is a typical case of weathering.

Exposure to the sun's rays of a coated product inside the car is a typical case of irradiation. The coating has no contact with the ambient climate. In this case, the surface of the irradiated coating is mostly raised to a higher temperature than the ambient temperature outside the vehicle. At the same time, the relative humidity immediately on the coating's surface is greatly decreased. The global radiation can also be affected as it passes through the car window. Atmospheric precipitation and natural air circulation are generally kept out. Outdoor exposure behind a transparent medium is generally called irradiation, as standardised in Method B of ISO 877 B. A typical example of irradiation is the use of xenon lamps as artificial radiation source, as found in the method for determining the hot colour fastness and the ageing behaviour of interior materials in motor cars according to DIN 75202. Table 2.1 compares weathering and irradiation by listing their essential features.

Table 2.1: Comparison of weathering and irradiation

Weathering stress factor	Climatic parameter	Weathering	Irradiation
Radiation	irradiance	controlled	controlled
Air oxygen	concentration	natural constant	natural constant (usually)
Heat	ambient temperature	controlled (independent on the irradiance)	non controlled (dependent on the irradiance)
Water	relative humidity	controlled (independent on the irradiance)	non controlled (dependent on the irradiance)
	rain amount, time of wetness	controlled	not included
Acid rain	special conditions	controlled, if occurring at "normal use"	not included
Air pollutions	concentration, action period	controlled, if occurring at "normal use"	controlled, if occurring at "normal use"
Purpose		weatherability	photo durability, (light fastness)

As found in the literature, weathering and irradiation are sometimes differentiated in terms of their purpose. Irradiation is commonly seen as a method for determining the **photodurability** or lightfastness. The purpose of weathering is to determine **weatherability**. Therefore, accelerated testing as a test method must always include the mutual action of global radiation (or its simulation by an artificial radiation source) and the other weathering stresses encountered in normal use. Because of the complexity of weathering stress and the synergy between individual stress factors, it can generally be stated that forgoing just one of the weathering stress factors found in normal use can jeopardise the transferability of the weathering results to practice. **Service life predictions for coatings for normal outdoor use can be expected to be false if derived only from irradiated specimens.**

Although the special importance of water as a weathering stress factor has been realized for the last ten years, it is for the most part not given due consideration in current standardised test methods. Generally, current, widely used test methods are still characterised by overly dry test conditions. Water must be given more consideration in accelerated testing to reflect real weathering stress. Watery precipitation, such as rain, dew, or fog, dictate the **time of wetness**, or TOW, along with the relative humidity. The water content of the exposed coating film, which has a great influence on coating ageing depends mostly on the TOW, as described in more detail in Chapter 3.2.3.

Figure 2.1 includes further environmental stress factors, such as acid precipitation (as rain or dew), aggressive dusts and aerosols, air pollutants and biological attack, e.g. by aphids, algae or fungi. They are less emphasised in this figure, because they are not found to have the same importance in all areas of normal use. However, they are known to boost the action of global radiation through synergy. In line with the aim of helping those engaged in coating testing, this book focuses on weathering as the key feature of accelerated testing.

2.2 Weathering versus corrosion testing

As mentioned in Chapter 2.1, weathering always includes actual or simulated global radiation in the test. Corrosion testing closely resembles weathering, but there are important differences. It is therefore appropriate to point out the distinction between **weathering** and **corrosion testing**, as used in this book. In paint chemistry, corrosion refers to destructive processes that predominantly occur on the surface beneath the coating, such as rusting of steel and filiform corrosion of aluminium. These are processes which the coating is meant to prevent. The essential differences between the two kinds of coating testing are compared in Table 2.2.

Table 2.2: Essential differences between weathering and corrosion testing

	Weathering	**Corrosion test**
Purpose	testing the ageing behaviour of coatings	testing the protection of the substrate against corrosion
Main criterions	1. decorative properties of the coating surface (optical properties appearance, intactness) 2. protection of the substrate (cracks- & craters, delamination)	protection for the substrate (degree of infiltration and rusting at a crack)
Coating	organic material	inorganic and organic material
Substrate	substrate-independent	metallic substrate
Main action factor	global radiation	corrosion promoting aggressively acting factors
Pre-damaging of the coating	no	defined cracks reaching the metallic substrate
Representative test standards	ISO 4892, ISO 877 (in laboratory) ISO 2810 (outdoors)	ISO 11997 (in laboratory), DIN 55665 (outdoors)

It is known that corrosion processes do not require the presence or action of global radiation. Therefore, the results of corrosion tests cannot be used to make predictions about the **ageing behaviour of coatings**. Furthermore, and in contrast to weathering, corrosion tests are not limited to organic coatings. They are also conducted on all other types of coating (organic and inorganic alike) whose primary function is to protect the substrate, for instance, sheet metal electroplated with nickel, chromium, or gold. Established corrosion tests employed in paint chemistry, such as determination of resistance to cyclic corrosion conditions of ISO 11997-1 [3], are predominantly used for making predictions as to how well a metal substrate can be protected by the test coating against corrosion caused by real or simulated outdoor conditions. In other words, corrosion tests are not directed at the ageing behaviour of the coatings involved. Since the action of global radiation is mostly excluded, corrosion tests cannot allow reliable predictions about the coatings' ageing behaviour. Furthermore, the test panels used are pre-damaged by deep scratches across nearly all of the specimen's surface, which complicates assessment of weather-induced changes. It was found by [5] that the changes in the appearance of coatings after subjection to **corrosion testing** (DIN 50018 [4]) in which a test atmosphere containing sulphur dioxide was employed did not correspond to the coating's ageing behaviour as determined by means of conventional outdoor weathering. Only one laboratory corrosion test, described in ISO 11997-2 [6], acknowledges global radiation as a stress factor and includes UV radiation in the test cycle. Another corrosion test, described in DIN 55665, is natural weathering of coatings. Determination of corrosion protection performance [7], published in 2007, automatically includes global radiation because the test is performed outdoors. In only one case was good ranking correlation reported between these two UV-inclusive corrosion tests. However, assessment of ageing behaviour remains complicated because of the pre-damage done to the specimen surfaces by scratches. Rusting and flaking near the scratch remain a criterion of assessment, even in these two test methods. **In summary, it can be stated that weathering cannot be replaced by corrosion testing.**

Table 2.2 lists further important differences between the two. Protection of the substrate is always in the foreground of corrosion testing. Changes in the decorative properties of the coating surface, such as loss of gloss, increased haze, yellowing, or scratch resistance, which characterise the coating's ageing behaviour, are of no interest. In contrast to corrosion testing, ageing behaviour is the most important result of weathering, irrespective of the substrate employed. Weathering can possibly lead to results of interest for assessing the protective function of a coating, too, for instance, when the coating surface reveals cracks, craters or delamination all the way to the substrate, opening up a path for substances such as oxygen, water, acid rain, and aggressive airborne pollutants that are most responsible for corrosion.

2.3 *Weatherability, service life, ageing behaviour*

The quality of a material is often generally described in terms of its **resistance** or **durability**. In the context of outdoor use, weather resistance, weathering stability, weather durability and weatherability are employed. All these terms mean the same. In other words, different authors use different terms to convey the same meaning. To avoid confusion, only the terms weathering resistance and weatherability are used in this book. In terms of coatings, weatherability characterises the ability to withstand the weathering stress that occurs in normal use.

In general, the reliability of a product is often expressed in terms of durability and functionality as well as breakdown or failure. The last two terms generally describe a state in which the product has lost its ability to function properly. The period of time which elapses until breakdown or failure occurs is normally used to describe longevity, a term used to describe quality and which is often called service life. However, when information about a coating's service life is sought, for instance, by looking at the product description, often what is found is that the coating is weather resistant, but no information is provided as to how long the product painted with that coating can resist the weathering stress encountered in normal outdoor use. This therefore raises the following question: Can the term weatherability sufficiently describe the coating's quality if no information about the time to failure is provided, i.e. without information about its service life? The answer is yes in the case of certain inorganic coatings. For instance, a surface coating made from gold, as mentioned before, can undoubtedly be viewed as infinitely weather resistant. In the case of organic coatings, though, the answer is no. The basis for these answers lies in the laws of thermodynamics. The question as whether or not a chemical reaction takes place depends on the value of the change in Gibbs' free energy, ΔG, which may be viewed as the reaction's momentum. ΔG is negative for a favourable reaction and positive for an unfavourable one. The degradation reaction which occurs between gold and oxygen to yield gold oxide is highly unfavourable because ΔG has a high positive value. Such a reaction will never occur in normal use because of the thermodynamics. The good states of preservation of gilded roofs from ancient buildings are testament to this fact.

The situation is totally different for polymeric materials. The oxidation reactions here are characterised by high negative values of ΔG. If it were only the laws of thermodynamics which applied, all polymers would spontaneously react with oxygen and water in normal use. However, thermodynamics only indicate the feasibility of a certain chemical reaction. In the case of polymers, the laws of reaction kinetics prevent spontaneous degradation.

Oxidation can only take place if the reagents receive a certain amount of energy, known as activation energy, from the outside. This activation energy can be introduced in the form of heat or radiation. In normal outdoor use, the highly energetic photons present in global radiation are able to provide this activation energy for the oxidation reaction. The number of sufficiently energetic photons absorbed by the polymer determines the number of macromolecules which are oxidised. This reaction commonly proceeds via a number of more or less stable intermediate products through to the final oxidation products carbon dioxide and water. Since the reaction requires global radiation for initiation, this kind of degradation is usually called photochemical ageing. Furthermore, as sufficiently energetic photons are always present outdoors in normal use, photochemical ageing of polymers cannot be prevented completely. All that can be done is to lower the rate of ageing by employing additives (pigments or stabilisers), as is described in more detail in Chapter 3.1.4. With progressive photochemical ageing, the coating properties also change for the worse at a higher or lower rate. **In other words, there is no such thing as weather-resistant polymeric materials, including organic coatings, since all organic materials are incapable of resisting weathering stress for an unlimited period of time.**

Therefore, breakdown of the coated product due to failure of the coating can always be expected. However, the term weatherability is often used without any information or qualification about the period of normal use during which the coating remains weather resistant, as determined, for instance, in ISO 2810 [8]. If accelerated testing, as based on that standard, still yields acceptable coating damage, then the term weatherability means that the tested coating resists the weathering stress only as defined in that standard. It would be wrong to apply the weatherability determined therein to normal use. When the assessment of the coating's weatherability is used without information about the expected service life, it can only be assumed that the coating is weather resistant for an appropriate service life. However, what is appropriate is always a matter of interpretation. In this context, a further question arises about the meaning of the term unacceptable coating damage in the assessment of the coating's weatherability.

Unacceptable coating damage is always a question of interpretation that also varies from case to case, as demonstrated in the following example. High-gloss coatings undergo gradual changes in their most essential quality characteristics, such as gloss and colour, with increasing exposure. Their service life can be adjudged to have ended when one or both of these ageing criteria have changed unacceptably. Considerable differences in the service life can be expected where

Table 2.3: Quality terms frequently used in combination with accelerated testing

1 **Quality term**	2 **Explanation**	3 **Basis**
Ageing behaviour	weathering caused changes of the coating properties	mutual action of the environmental action factors during weathering
Service life (actual, estimated or appropriate)	period till failure of the coating at "normal use", (measure of the longevity)	ageing behaviour + "normal use"
Weatherability	no failure till the defined or appropriate (undefined) service life at "normal use"	ageing behaviour + service life at "normal use"
Guaranty period	mostly gratuitously defined minimum service life	actual, estimated or appropriate service life

identical coatings are used for luxury cars and for tractors. All these examples clearly show the incorrectness of the term weatherability. Properly, each accelerated testing only lend itself to predictions about the ageing behaviour of the coating to be tested. **Therefore, coatings should be predominantly judged on the basis of their ageing behaviour.** From knowledge of the ageing behaviour, both weatherability and service life can be derived as a function of the coating properties chosen to serve as ageing criteria, the product-dependent permissible changes, and the real weathering stress expected in normal use. The most important quality terms in the context of accelerated testing are listed in Table 2.3.

Unfortunately, the term **weatherability** is firmly established in the literature, including the standards. To minimise the number of terms and to avoid confusion, weatherability is used in this book despite its incorrectness in referring only to an undefined, appropriate service life. It is now time to discuss some further essential aspects of the term **service life**. In spite of different interpretations, a coating has always reached the end of its service life when, as a result of cracks, blisters, craters or flaking, substrate-damaging stress factors, such as oxygen, water, and airborne pollutants, gain unhindered access to the substrate. The end of the service life has also been reached when the surface of the coating has changed in gloss or colour or lost its intactness to the extent that the coated product no longer fulfils the aesthetic or other user-related requirements. Presently, paint makers and coating users alike feel increasingly compelled to grant a defined minimum service life in the interests of consumer protection. The actual attainable service life for normal use should exceed the guaranteed one to an economically justifiable extent. As mentioned before, the end of the service life is often defined subjectively on the basis of what is deemed appropriate. This might be the type of coated product, the geographical location of its use, the present market valuation, and the normal use agreed by paint makers and coating users alike. For instance, permanent use in industrial conurbations, daily roadside parking or beneath trees, regular use of commercial car washes currently constitute normal use for cars. The definition of normal use often develops into a source of dispute, especially where complaints are concerned. However, there is also no disputing the fact that a coating cannot and need not resist all weathering stresses, especially extremes that seldom occur. For this reason, guarantees of minimum service life are often restricted to use in certain geographical regions. For instance, coatings for cars that are used only in Central Europe need not fulfil the more stringent requirements for use in subtropical industrial areas such as Jacksonville and Boston in Southern USA. This is important, because coating of the car body contributes a substantial amount to a car's purchase price.

Where a service life is guaranteed, the quality term weatherability means no reason for complaint until the end of a period equal to at least normal use. In the not inconsiderable DIY market, for instance, it is difficult to find any such data, although here, too, the purchaser wishes to obtain information about the service life of the coating he intends to buy and use. He expects the service life to reflect the present state of the art in paint chemistry. However, the expected and the real service life often fail to match. Let us stay with the DIY example. It can be safely assumed that nearly all readers of this book have had bad experiences when it comes to the service life of a coating, for instance, in their own dwellings or leisure pursuits. If the paintwork of garden furniture that you lovingly coated yourself starts flaking after a short summer in use, there would certainly be cause for complaint. To back up your case, you would cite the declaration on the container that the product is suitable for indoor and outdoor use, and perhaps the expiry date for the paint in the unopened container, but you will not find information about the coating's service life in defined normal use. In other words, there is little hope of winning your case from a guarantee viewpoint. This is annoying, but is still relatively innocuous.

However, complaints about poor coating quality can quickly arise in the automotive industry, with claims for damages reaching several million Euros. In order to avoid this, such coating users will insist on as much reliable evidence as possible that the delivered cars will retain their initial attractive and intact appearance over the guaranteed service life during use in that country and intended export countries. If the service life expires within the term of the guarantee, complaints by the car owners will be unavoidable and might impact on further sales. Increasing competitive pressure along with increasing economic and environmental restrictions, especially in the auto market, are making it necessary to couch failure criteria more closely. Conversely, overly high sureties, where they exist, are becoming unaffordable. **In other words, expectations about the transferability of the results of accelerated testing to actual practice are steadily rising.** Characterising a coating's quality simply in terms of weatherability, without further definition, will prove inadequate for meeting rising requirements in the future! Determining a coating's ageing behaviour on the basis of accelerated testing leads to more reliable predictions of service life.

2.4 Accelerated testing and service-life prediction

As already discussed in Chapter 2.3, knowledge of the ageing behaviour and a reliable service life prediction derived from that are gaining in prominence in coating technology. In this context, the question arises as to which possibilities paint makers and coating users have of acquiring this know-how. Whereas paint makers have extensive theoretical means due to their training practical experience, coatings users insist on practical evidence so as to be able to estimate a coating's aging behaviour more or less exactly in order be sure of avoiding expensive complaints. Let us stay with the example of the automotive industry. In this scenario, a car manufacturer offers a 5-year guarantee on the car body finish. Therefore, the easiest way to test this promise would be to drive appropriate cars under conditions of normal use prior to series production. However, the competitive situation mentioned above renders such a method of demonstrating the coating's service life virtually unaffordable now. A weathering test of that kind is time-consuming and cost-intensive. To ensure adequate statistical certainty, at least ten cars would have to be subjected to each type of normal use. In other words, a lot of cars would have to be involved in the test. The resultant time lapse between the coating's creation and its implementation in practice would be unacceptable, especially in light of the speed at which coatings are developing. Therefore, there are only a few cases of normal use tests being conducted under actual weathering stress.

The information necessary for effectively and reliably predicting the coating's service life is provided by accelerated testing. However the current, mostly standardised methods of accelerated testing do not always fulfil this task adequately. In many cases, they are not even up to date. **Accelerated testing itself must be continuously developed to reflect advances in coatings.** New creations by the coating industry often require new test method designs, for instance, that change the weighting of the acting weathering stress factors. A further goal of this book will be to provide the reader with the tools for critically examining methods of accelerated testing. As will be discussed later in Chapter 8.3, test standards always lag behind advances in coatings.

2.5 *Accelerated testing – time-compressed or accelerated*

The desire to acquire evidence of the service life that can be expected in normal use as quickly as possible is as old as coatings development itself. Its fulfilment has remained a challenge for coatings testing down to the present. Accelerated testing was the test method arrived at for meeting it. According to the statements made above, accelerated testing can be expected to be successful only if the weathering stress factors leading to coating ageing are known and simulated in such a way that coatings are aged at higher speeds under test conditions than in normal use, but without unacceptable loss of practice correlation. Until now, it was frequently not known which of the weather conditions outdoors needed to be involved or not in the weathering stress and designed into the weathering test method. A look at current standardised methods of accelerated testing reveals that major stress factors, such as moisture, are underestimated while others are given undue attention, such as phases of darkness whose contribution to coating ageing are questionable. According to VDI Guideline 3958-1 Environmental Simulation, Fundamentals and Methods [9], attempts to shorten the exposure period are nowadays increasingly based on worst case weather scenarios. First, it needs to be established what is meant by a worst case weather scenario. It describes a real weather situation that causes the coating to age particularly rapidly or at least faster than in other situations. It generally makes the greatest contribution to ageing. It is characterised by a regular simultaneous occurrence of weathering stresses that are particularly destructive and whose combined effects are often boosted by synergy.

A selection of typical worst case weather scenarios is listed in Table 2.4. They all are weather situations characterised by the action of sunshine in specific combination with one or more of the other environmental stress factors, such as heat (high temperature), moisture (high or low relative humidity, rain), and acid atmospheric precipitation (acid rain, dew, or fog). As can be seen, sunshine is part of all worst case weather scenarios. Different types of coatings can have different worst case weather scenarios. Furthermore, different worst case weather scenarios are also possible if different ageing criteria are applied. This can be explained, for instance, with pigmented coatings. Resin and pigments are degraded by different ageing mechanisms. It is possible, for example that the loss of gloss caused by preferential degradation of the resin and the change in colour caused by preferential degradation of the pigment can lead to competing worst case weather scenarios. An example taken from the literature [10] (more detail is provided in Table 3.9) demonstrates how the type of worst case weather scenario can depend on the chosen ageing criterion. In this example, the results of two outdoor weathering tests performed on identical series of pigmented coatings out in Miami, Florida and Aquaba, Jordan, were compared. Where loss of gloss was the ageing criterion, the test carried out in the hot dry desert climate of Aquaba, Jordan, revealed the fastest coating ageing. In contrast, where chalking was the ageing criterion, the humid subtropical climate of Miami, Florida, yielded the fastest coating ageing. Including a dry-heat weather situation as the possible worst case weather scenario in Table 2.4 means that the stress caused by a hot desert climate can also lead to the highest rate of ageing. Consequently, high moisture, which is a weathering stress factor that usually accelerates ageing, can also slow down the negative changes of certain coating properties during weathering. However, such cases are seldom and mostly limited to one of the ageing criteria.

Table 2.4: Possible worst case weather scenarios capable of affecting coatings in normal use

Weather scenarios
Sunshine at high temperature and high relative humidity (humid-hot)
Sunshine after longstanding rainfalls
Sunshine at high temperature and low relative humidity (hot-dray)
Sunshine after acid rain or acid dew

Table 2.5: Time-saving strategies in accelerated testing

	Time compressed test	Accelerated test
Choice of the test parameters	without significantly exceeding of the extreme values of the actual weathering stresses	with significantly exceeding of the extreme values of one or several of the actual weathering stresses
Recommended test method	simulation of the "worst case" weather scenario (if known) with higher frequency than encountered during normal use	simulation of the "worst case" weather scenario (if known), similar to the time compressed test, but using test parameters with increased severity
Performance	natural and artificial weathering	predominantly artificial weathering

As described in VDI Guideline 3958-1 Environmental Simulation, Fundamentals and Methods [9], accelerated testing can be designed as time-compressed or accelerated test methods, which are common time-saving strategies in environmental testing. The essential differences between these two types of simulated weathering stress are listed in Table 2.5. Where the **worst case weather scenario** or scenarios of the coatings or coating series to be tested are known, accelerated testing is designed as a time-compressed method by including the respective worst case weather scenario or scenarios in the test cycle, but much more frequently than they occur during normal use. It is this higher frequency which makes the time-saving effect possible. **An essential feature of time-compressed accelerated testing is that the chosen test parameters do not or only slightly exceed the extremes of real conditions of normal use.** This method ensures that no ageing mechanisms are included which would never occur in normal use. Avoiding such unnatural mechanisms also ensures that service-life prediction based on time-compressed methods can be seen as more reliable than those based on the accelerated method discussed below.

The accelerated test method employs test parameters that considerably exceed the real conditions of normal use. One more for good measure is the underlying maxim by which coating ageing is expected to accelerate in line with the degree to which the test parameters are exceeded. The aim of accelerated test methods is to considerably shorten the necessary exposure period relative to the time-compressed methods. A typical feature of accelerated methods is the acceptance that the unrealistically exceeded test parameters may initiate unnatural ageing mechanisms that yield a loss of correlation with reality. It can make sense to use worst case weather scenarios too for designing an accelerated testing test in which one or more of the test parameters, such as photon energy, radiation, content of airborne pollutants, and acidity of rain, is exaggerated.

Instances of such accelerated testing tests are known that needed only a fifth of the time needed by time-compressed method to produce the same ageing effect. Certain accelerated testing tests have justified their right to exist, especially for the monitoring of series production. Accelerated test methods cannot be recommended for testing new coating creations based on new raw materials, formulations or types of application. The outcome of over-acceleration might be that new coating creations which are excellent for normal use could be adjudged unsuitable and therefore prevented from going into series production. **Nowadays, time-compressed and moderately accelerated testing tests are frequently used in combination.**

Special weathering test methods, known as **ultra-accelerated testing**, have been published in the literature. These methods expose the samples to radiation of unnaturally short wavelength, extremely high-intensity radiation from UV excimer laser, or special lamps with plasma discharges in highly reactive gas mixtures and others. According to the inventors, these methods enable service-life predictions to be made after just a few hours' exposure. According to the present state of the science, such types of ultra-accelerated testing are completely devoid of reality, and must be considered as silly nonsense. The reader should agree with this judgement after reading Chapter 3. In contrast, methods that work with shortened exposure periods and performed with common test parameters combined with highly sensitive analytical methods constitute a further useful and modern time-saving strategy that offers an alternative to the conventional types of accelerated testing. Those methods employ common weathering techniques, but do not apply the common ageing criteria, such as loss of gloss and colour changes. They use such special analytical techniques as IR spectroscopy, chemoluminescence, and thermoluminescence. These analytical techniques enable ageing-related changes in the coating to be identified long before the common ageing criteria become visible. There are reports that the ageing-related differentiation found in a series of different automotive coatings samples exposed for 40 days were found in the same weathering test after just 10 days of exposure to chemoluminescence. However, more work is needed on detecting such early-occurring, invisible ageing effects reliably and at an acceptable cost. These efforts can be seen as a means for further shortening the duration of accelerated testing without loss of correlation with practice.

Real weathering stress, whether time-compressed or accelerated, can be simulated in both natural or artificial ways. The natural form of accelerated testing is **outdoor weathering** under natural weathering stress. The essential feature of all outdoor weathering tests is the use of the sun as the primary source of radiation. For accelerated testing in laboratory weathering devices, known as artificial weathering, corresponding devices and artificial radiation sources are used to simulate weathering stress in normal use.

3 Effects of weathering stress

3.1 Coating ageing by weathering

3.1.1 General aspects

Weathering stress on coatings generally manifests itself in ageing of the coating. In order to understand the processes leading to coating ageing better, the mechanisms that occur in the coating material during weathering will be described in more detail. The essential weathering stress factors shown in Figure 2.1 are discussed one after the other in this book. In general, ageing of materials is, according to DIN 50035 [11], the totality of all chemical and physical processes that occur irreversibly in a material in the presence or absence of external influences. Outdoors, external influences may be regarded as the dominant influences. These influences and their causes are listed in Table 3.1.

Undoubtedly, the primary reason for coating ageing in outdoor use is that of weathering stress. Therefore, all aspects of accelerated testing can be limited to weathering stress characterised by all stress factors acting thermally, chemically, mechanically, and biologically during outdoor use, whether alone or in combination. The quality of accelerated testing can be judged by the ability to simulate the complexity of weathering stress. There are examples that prove that certain stress factors are boosted enormously in combination due to synergies. Tests on automotive coatings, for instance, have revealed that attacks by acid atmospheric precipitation or by biological substances only led to considerable damage during artificial weathering if these stress factors were combined with others, such as global radiation, heat and water.

Furthermore, coating ageing can be differentiated as a function of ageing reaction. Table 3.2 shows a listing of known weather-induced ageing reactions that lead to irreversible chemical alterations in polymeric materials due to outdoor weathering. Due to the predominant role of radiation during weathering, these reactions are generally called "photochemical aging" or "photodegradation" .

These reactions occur in different ratios simultaneously or consecutively during weathering. The degradation of PVC is a good example of a complex photodissociation reaction (here: dehydrochlorination and photooxidation). Since normal outdoor use of coatings includes the presence of radiation, as well as atmospheric oxygen, photooxidation can be seen as predominant ageing reaction.

Table 3.1: Reasons and major effects on coating ageing during outdoor weathering

Cause		Essential effect
1	thermal stress	thermal degradation
2	chemical stress	chemical degradation
3	mechanical stress	molecule scission
4	stress by ionicing radiation	radiation-caused chain breaking or branching (cross linking)
5	biological stress	biological degradation (e.g. composting)
6	stress by weathering (mutual action of the stresses from the natural environmental)	photochemical ageing

Ulrich Schulz: Accelerated Testing:
Nature and Artificial Weathering in the Coatings Industry

ISBN: 978-3-86630-908-1

As already discussed in Chapter 2.3, carbon dioxide and water are the stable end products of the photooxidation of all polymeric materials according to the laws of thermodynamics. The reader may recall that the energy threshold, the so-called activation energy, must be overcome before photooxidation can occur! Global radiation contains photons of sufficient energy for this purpose. However, the polymer must absorb these photons in order to transfer their energy to polymer molecules. Therefore, absorption of radiation is always the primary photophysical reaction of the ageing process. If the absorbed photon has enough energy, it may be able to activate the affected polymer molecule and promote it to an excited energy state. The excited molecule is no longer within its energy equilibrium and therefore strives to get rid of its surplus energy as fast as possible using one of the deactivation processes listed in Table 3.3.

Which of the three deactivation routes is chosen by an excited polymer molecule predominantly depends on the photon energy and the ambient conditions, as described later in Chapter 3.1.3. If activation caused by absorption of a photon does not initiate photooxidation, deactivation will predominantly occur intramolecularly by physical processes that cause no damage to the polymer. Those processes which take place inside the polymer molecule and usually proceed over several steps are described in more detail in [13] and [14]. For our purposes, we shall explain the key concepts using a simplified form of the so-called Jablonski diagram shown in Figure 3.1.

This schematically shows activation and deactivation for the energy states S_0, S_1 und S_2 (singlet states) und T_1 und T_2 (triplet states). The thick horizontal lines above and below the energy levels

Table 3.2: Essential reactions causing photodegradation (adapted from [12])

Weather-caused degradation (photochemical ageing)	
1	photooxidation
2	photodissociation (e.g. dehydrochlorination of PVC)
3	peroxyde and hydroperoxyde conversion
4	Norrish reactions Type I and II
5	photo-Fries reaction
6	photosubstitution
7	photoaddition
8	photoelimination
8	phodimerization
9	photocondensation
10	photoisomerisation
11	photoionisation

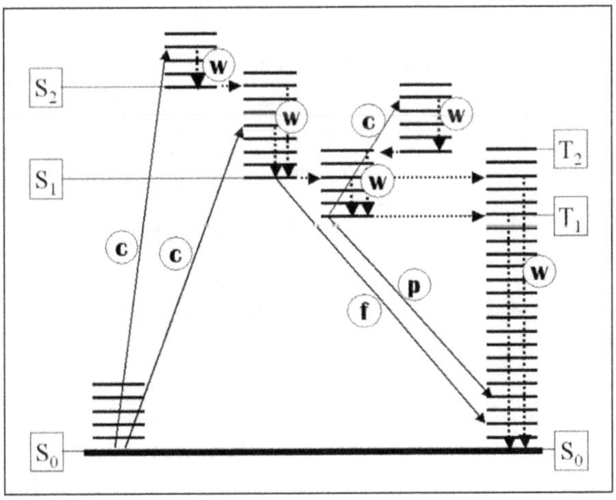

Figure 3.1: Energy states of organic molecules ("The Jablonski diagram")

Table 3.3: Deactivation processes of excited polymer molecules

Deactivation process		Kind of energy transfer
1	intramolecular deactivation	fluorescence (f), phosphorescence (p) warming (w)
2	intermolecular deactivation	energy transfer to an acceptor (e.g. a stabilizer)
3	deactivation by chemical reaction	breaking of chemical bonds causing reaction with oxygen leading to chemically changed products

characterise the respective vibration or rotation levels, but are not discussed in more detail. The sloping arrows © mean absorption of photon energy. The dashed horizontal arrows stand for radiation-free energy transfer. The downward arrows (f and p) indicate intra-molecular deactivation. When a photon is absorbed ©, the accepting polymer molecule is promoted from its base energy level S_0 to one of the excited energy levels S_1, S_2, or T_2, the choice of level depending on the photon's energy. Intramolecular deactivation proceeds by emission of radiation in the form of fluorescence (f) or phosphorescence (p) from the singlet state levels S_1 and S_2 or from the triplet state level T_2 back to the base energy state level S_0 or by slight heating (w) of the polymer.

In the case of intermolecular deactivation, the excited polymer molecule donates its surplus energy to another. If the accepting polymer molecule can dissipate this energy in a harmless form, such as heat or infrared radiation, no chemical alteration to the polymer molecule will happen. Weathering-induced damage of polymers can only occur if the latter type of deactivation by chemical reaction takes place. In this case, the energy of the absorbed photon serves as the activation energy for initiating photooxidation of the excited polymer molecule that leads to its degradation.

Start:	$RH \xrightarrow[k_1]{h\nu} R\cdot + \text{Initiator-H}$	(1)
Propagation:	$R\cdot + O_2 \xrightarrow[k_2]{} ROO\cdot$	(2)
	$ROO\cdot + RH \xrightarrow[k_3]{} ROOH + R\cdot$	(3)
Branching:	$ROOH \xrightarrow[k_4]{h\nu} RO\cdot + OH\cdot$	(4)
Termination:	$R\cdot + R\cdot \xrightarrow[k_5]{} R\text{-}R$	(5)
	$ROO\cdot + R\cdot \xrightarrow[k_6]{} ROOR$	(6)
	$ROO\cdot + ROO\cdot \xrightarrow[k_7]{} ROOR + O_2$	(7)
Catalytic effect of metals (example Fe):	$2\,ROOH \xrightarrow{h\nu + Fe^{2+/3+}} ROO\cdot + RO\cdot + H_2O$	(8)

Figure 3.2: Mechanism of photooxidation (adapted from [15] and [16])

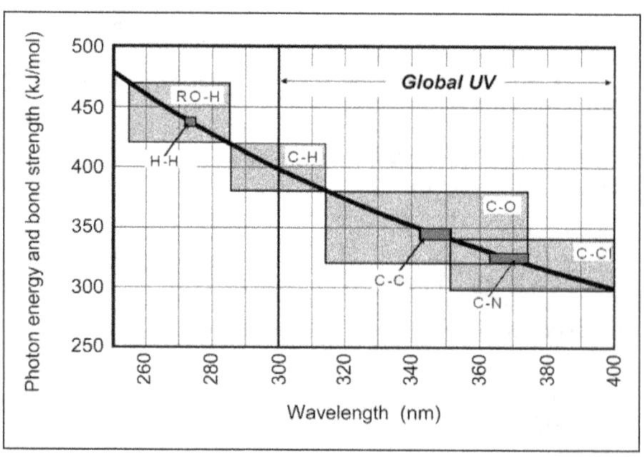

Figure 3.3: Comparison of photon energy and bond strength (adapted from [17])

3.1.2 *Photooxidation – predominant type of coating ageing*

This section of Chapter 3 is devoted to photooxidation as the predominant type of ageing undergone by coatings in normal use. The mechanism by which activation and oxidation of polymer molecules starts and progresses is demonstrated for polyethylene in Figure 3.2. This example is often used to simplify the explanation. Photooxidation of coatings proceeds similarly, but is much more complicated.

In general, oxidation of a polymer molecule proceeds in the form of a free-radical chain mechanism. It starts, as shown in Figure 3.2, with absorption of photon energy "hν" to generate a polymer free-radical "R·". The initiation reaction (1) is triggered by the energy-rich photons of the global radiation. To trigger the reaction (1), the photons must be capable of breaking the polymer bond. Therefore, photon energy and bond strength must at least be equal. The black curve in Figure 3.3 shows the photon energy as a function of the radiation's wavelength. The shaded rectangles indicate the range of bond strengths of the polymer bonds concerned as a function of the wavelength of the radiation. This chart illustrates the wavelength range over which the radiation can break up a specific type of polymer bond.

In line with the detailed explanations of the spectral distribution which follow in Chapter 3.2.1.2, the global radiation contains radiation down to a wavelength of 300 nm, equivalent to a photon energy of at least 398 kJ/mol. Global radiation is therefore able to cleave the polymer bonds most frequently found in common coating resins, and to initiate the photooxidation. Furthermore, the position of the vertical right edges of the shaded rectangles shown in this figure indicate that only that part of global radiation which has a wavelength of less than 400 nm is able to cleave the usual polymer bonds. The part of global radiation that is dangerous for polymers is called global UV.

The free-radicals generated by the free-radical chain mechanism of photooxidation shown in Figure 3.2 are highly reactive because of free valencies. They are predominantly available for reaction with oxygen. The free-radical R· generated in the initiation reaction (1) reacts with oxygen to form the free-radical ROO· as show in in reaction (2). This is followed by reaction (3) which generates hydroperoxide (ROOH) und a further free-radical R·. Since both of the so-called propagation reactions (2) and (3) also lead to free-radicals, the mechanism of photooxidation is able to occur without a further reaction initiated by an absorbed photon. In other words, once initiated, photooxidation could be maintained without further irradiation if it were not terminated by other reactions. Theoretically, one initiation could lead to the complete destruction of the polymer. The so-called branching reaction (4) has an accelerating effect on polymer degradation. It takes place when the hydroperoxide generated by reaction (3) absorbs a further photon h**ν**. Propagation reaction (4) forms two new polymer free-radicals RO· and OH· without consuming another one. Reaction (4) can lead to branching of the free-radical chain mechanism. The scission of hydroperoxide can be catalytically accelerated by metal ions capable of existing in several valencies, represented here in Figure 3.2 by reaction (8) with the example of iron. The marked variability in the service life of the PMMA used to roof the buildings of the Olympic Park in Munich in 1972 is a good example of the acceleration of polymer degradation due to iron. A few parts per million of iron inadvertently got inside the material during manufacturing, causing a difference in service life of about 30 years [18].

If the free-radicals do not quickly find the oxygen which they need by way of reaction partner, the photooxidation mechanism can be terminated by recombination of free-radicals, as represented by reactions (5) (6), and (7). Recombination of free-radicals leads to slight warming, frequently combined with formation of stable intermediate products (R-R), such as ketones and alcohols, represented by reaction (5). If termination is due to recombination of free-radicals belonging to the same polymer molecule, chemical change will mostly not occur. However, recombination of free-radicals from different polymer molecules can lead to crosslinking, and is mostly accompanied by increased brittleness. Reactions (6) and (7) lead to the formation of carbonyl groups (ROOR). As will be described later, carbonyl groups are so-called chromophores that enormously contribute to the absorption of UV radiation. Intermediate products generated by termination of the photooxidation can be degraded by further photooxidation and the same mechanism as far as the stable end products, namely carbon dioxide and water.

It is known that highly pure polymers have very good weatherability. The reason is that pure polymers themselves are unable to absorb the radiation which has the necessary photon energy for breaking chemical bonds. The mechanism of photooxidation cannot be triggered without absorption of radiation. However, technical products, including standard coatings, mostly contain enough additives and impurities acting as chromophores that are able to absorb photons and transfer their energy to the

Table 3.4: Extinction coefficient and quantum efficiency of two typical chromophores (from [17])

Chromophore	Extinction (at 310 nm) Mol^{-1} cm^{-1}	Quantum efficiciency
Hydroperoxides	0.4	1
Carbonyl groups	10	0.01 to 0.08

polymer molecules. The most important chromophores are hydroperoxide and carbonyl groups. They can be formed during weathering, as described earlier. However, they can already exist in the virgin unweathered polymer, for instance, having been formed by processes during the polymer's prehistory. Table 3.4 shows the essential properties of both these chromophores.

The extinction coefficient characterises the ability to absorb UV radiation, in our case radiation of 310 nm. Quantum efficiency indicates how many of the photons absorbed by the polymer lead to the formation of polymer free-radicals as a prerequisite for photooxidation. In comparison, hydroperoxides are less able to absorb UV; however, almost each of the absorbed photons leads to free radicals. Carbonyl groups are better able to absorb UV, but the number of absorbed photons leading to free-radicals is lower. For all the chromophores inside a polymer, the quantum efficiency will only be 10^{-2} to 10^{-5}, i.e. of the 100 to 100,000 polymer molecules that are hit by a UV photon, only one leads to a chemical change. Whether excitation of a polymer molecule by a UV photon leads to photooxidation or whether the surplus energy is dissipated by one of the physical types of deactivation depends on a large number of internal and external conditions, as will be discussed in the next chapter.

3.1.3 Influences on photooxidation

Figure 3.4 shows the key influences on the two basic reactions - activation and deactivation - of polymers outdoors in normal use.

Activation of a polymer molecule (represented here by the black horizontal arrow from A (S_0) to A ($S_{1;2}$)) is predominantly influenced by the quantity and quality of the available global UV (radiation EUV and wavelength λ_{UV}), as well as by the type of coating itself. The coating predominantly influences activation of the molecule via its content of chromophores and light stabilisers, as well as its formulation and prehistory. The environmental stresses listed on the right of Figure 3.4 are responsible for the path of deactivation. They determine whether deactivation of excited molecule A takes place by chemical reaction with oxygen to form a stable new molecule B (represented by the black horizontal arrow from A ($S_{1;2}$) to B (S_0)) or by a physical type of deactivation to the base energy state A (S_0) without change in molecule A (the returning grey arrow from A ($S_{1;2}$) to A (S_0)). Only the path from A (S_0) to B (S_0) is of interest in accelerated testing. The key prerequisite for the formation of oxidation product B is the presence of oxygen, of course. That does not mean only the oxygen in the ambient atmosphere.

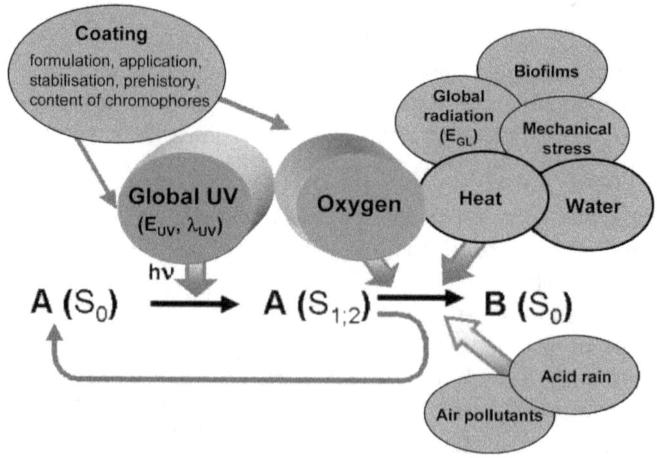

Figure 3.4: The key influences on activation and deactivation

More important is the presence of oxygen directly where the free-radicals are formed. The coating affects this process by its ability to dissolve and transport oxygen, expressed by the rate of oxygen diffusion inside the coating. A further key prerequisite for oxidation of a polymer is its content of free-radicals available for reaction with oxygen. The content of free-radicals is heavily influenced by the presence of light stabilisers, especially free-radical scavengers or antioxidants that are able to avoid or deactivate free-radicals before meeting an oxygen molecule. The type of coating, especially its major polymer bonds, also influences the probability that the deactivation follows the path of photooxidation. Of course, the other environmental influences shown in Figure 3.4, such as heat, water, and global radiation, are key factors in the probability of whether photooxidation occurs or not. In the event that photooxidation occurs, the rates of the reactions leading to polymer ageing are directly related to the service life. The rate of photooxidation (v_{ox}) determines the rate of polymer ageing. The relationship between v_{ox} and the weathering stress factors listed in Figure 3.4 can be explained by means of the highly simplified equation for polyethylene, which is shown in Figure 3.5.

$$v_{ox} = \frac{E_e^{0.5} \cdot k_2 \cdot k_3 \cdot c_{O_2} \cdot c_{RH}}{k_2 \cdot k_7^{0.5} \cdot c_{O_2} + \cdot k_3 \cdot k_5^{0.5} \cdot c_{RH}}$$

Figure 3.5: Influences on the rate of photooxidation as exemplified by PE (adapted from [15])

The k-values in the equation are identical with those for the free-radical chain mechanism shown in Figure 3.2. Each of the reactions listed in Figure 3.2 has its own reaction rate coefficient, labelled k_1 to k_7. **It is a law of reaction kinetics that the slowest reaction step determines the rate at which a complex process proceeds.** Therefore, the initiation reaction ((1) in Figure 3.2), which is known to occur very quickly, can be neglected in favour of the slower chemical reactions (2) to (7). From experience, only k_2, k_3, k_5 and k_7 need be considered. However, the reaction rate coefficients, which are characteristic of a specific chemical reaction, are not responsible on their own for the rate of reaction. It is a further law of reaction kinetics that the rate of reaction is determined by the concentration of the reaction partners at the point where the reaction takes place. The concentration of cleavable polymer bonds is expressed by c_{RH} und the concentration of oxygen by c_{O2}. The content of polymer free-radicals is predominantly determined by the UV radiation E_e, which in this example has an exponent of 0.5. For simplicity, it is admissible to regard just the numerator of the equation and to ignore the denominator. Thus, it can be seen that photooxidation does not take place without irradiation (expressed by E_e). Reactions (2) and (3) are the most important (expressed by k_2 and k_3). Photooxidation also requires the presence of oxygen (expressed by c_{O2}), and cleavable polymer bonds expressed by c_{RH}. The value of c_{RH} can change as a function of the photon energy of the acting radiation. In other words, the higher the photon energy, the more cleavable are the stable polymer bonds. These explanations show that all influences on the ageing of coatings that are due to weathering stress and polymer properties bear a simple relation to the equation in Figure 3.5.

3.1.4 Efficiency of light stabilisers

In normal test practice, it is rare for coatings tested not to contain additives for providing a degree of protection against weathering stress. Determining the efficiency of such protective coating additives are often the task of accelerated testing. It thus makes sense to pay particular attention to these additives. In this book, all coating additives whose task is to prevent or at least substantially retard coating ageing are included under the term “light stabilisers”, irrespective of their type and mode of action. Light stabilisers can only fulfill their protective function if they themselves have adequate weatherability. The following text deals with the most important and commonest types of light stabiliser, and especially with their influence on the process of photooxidation. More detailed explanations can be found in [16].

From Figure 3.2, the simplest possibility of protection might be to prevent the photons of global radiation from entering the polymer. That would prevent the free-radical chain mechanism of photooxidation from starting. This is the principle employed in the oldest form of light protection in which pigments reflect or absorb harmful radiation before making contact with polymer molecules. Titanium dioxide and carbon black are the most extensively studied pigments frequently used for polymer materials. They are known for their high capacity to absorb UV radiation. As shown in Figure 3.6, the absorbance of the titanium dioxide modifications anatase and rutile increases dramatically as the wavelength falls below 450 nm. Both TiO_2 modifications show very good absorbance at the most destructive wavelength of global radiation of around 320 nm, as discussed in more detail in Chapter 3.2.1.2.

Due to their inherent strong colour, however, pigments cannot be used in every coating. Furthermore, certain pigments sometimes cannot be used without special precautions, due to their photoactivity. In particular, titanium dioxide is known to accelerate coating ageing. Figure 3.7 demonstrates this for an alkyd-resin-based coating pigmented with anatase. The Figure shows the rate of coating ageing (as the rate of resin degradation) as a function of the content of TiO_2 upon irradiation with UV of wavelength 320 nm. As may be seen, coating ageing is accelerated with increasing content of TiO_2, although a decreasing ageing effect would have been expected due to increasing UV absorbance. Therefore, TiO_2-particles intended for pigmenting coatings in outdoor use need to be treated beforehand in order that their photoactivity may be adequately reduced.

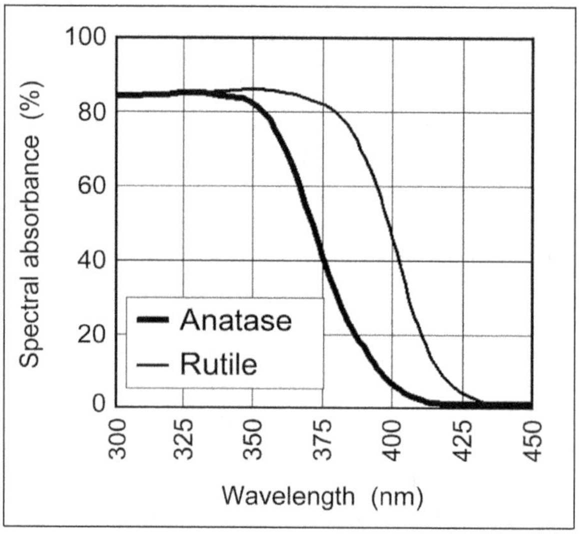

Figure 3.6: Spectral absorbance of pressed chips of different types of titanium dioxide (from [19])

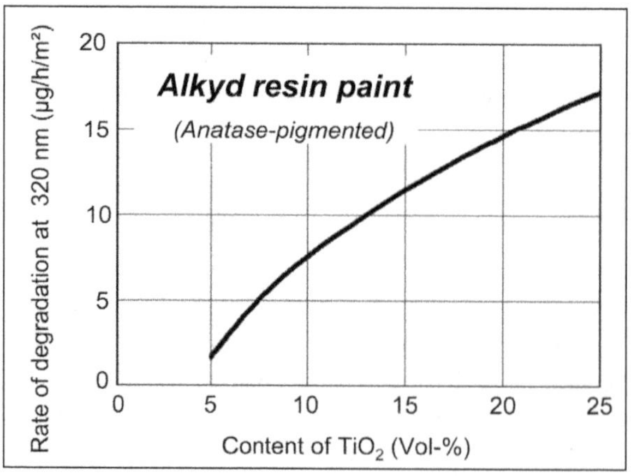

Figure 3.7: Rate of degradation during weathering of an anatase-pigmented film of an alkyd resin paint as a function of the content of titanium dioxide (from [20])

However, absolute UV protection is not possible, not even in the case of highly pigmented topcoats, such as aircraft coatings. Pure pigment particles, as well as small areas of pure resin, which are exposed directly and unprotected to weathering stress factors can be found inside the outermost micrometers of the coating surface. As far as the resin is concerned, only those areas which are beneath the pigment particles are protected. On the other hand, exposed pigment particles are not protected against critical weathering stresses, such as acid rain and aggressive air pollutants. Therefore, pigments are unable to adequately prevent weather-induced coating ageing of the outermost micrometers. To an extent depending on their individual weatherability, either the pigment or the resin is predominantly degraded in outdoor use.

Figure 3.8 shows the different types of weathering-induced surface damage that can be found in pigmented coatings. The upper picture shows the unweathered state. Free parts of the pigment particles (black circles) as well as free resin areas (grey area) can be seen in direct contact with the environment (white area). The two lower pictures depict two possible states after weathering. The lower left picture shows the result of mostly resin degradation. Undamaged pigment particles are released from the resin and are easily removed on contact with other media in a process usually called chalking. The lower right picture shows the outcome of mostly pigment degradation. The pigments are invisibly destroyed and the residual gaps in the resin lead to more extensive surface roughness and loss of gloss or greying.

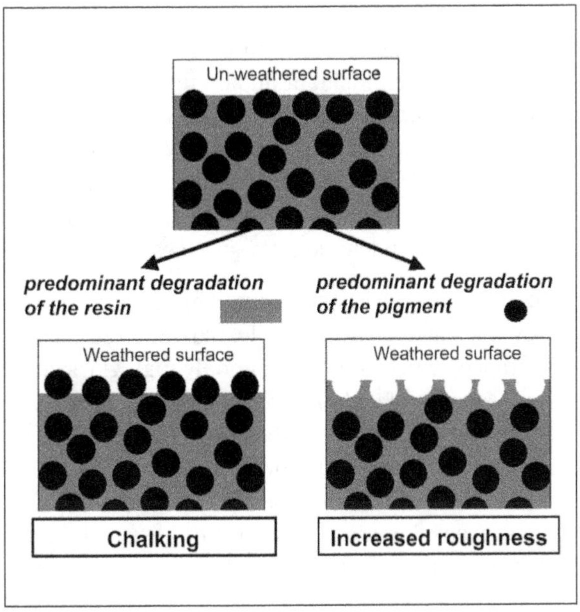

Figure 3.8: Two key types of weathering-induced surface degradation of pigmented coatings

With certain coatings, heterogeneities in the pigment distribution that can occur during drying or curing of the coating take the form of Bénard cells and can lead to visible differences in the rate of coating ageing on the coating surface. Figure 3.9 shows a typical structure on the surface of a PUR-based aircraft coating after two years of outdoor exposure in Leipzig, Germany, due to Bénard cells.

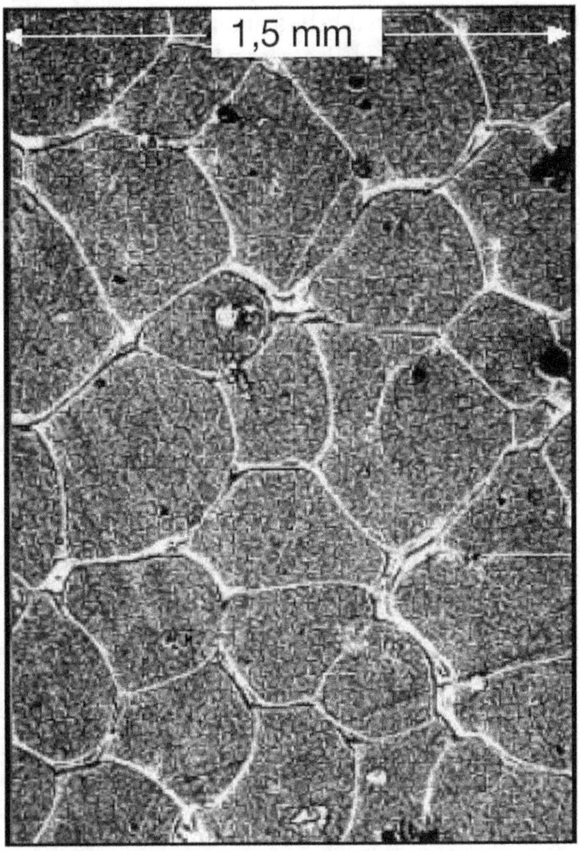

Figure 3.9: Bénard cells on the surface of a PUR aircraft coating after two years of outdoor weathering in the urban environment of Leipzig, Germany (from [21])

Deficiences during application (e.g. non-optimal viscosity, density or homogeneity of the wet coating) can lead to completely pigment-free surface layers that are several micrometers thick and totally unprotected against weathering stress. Such a surface can exhibit the same poor weatherability as the corresponding non-pigmented coating.

On account of their intrinsic colour, pigments are not used for clearcoats. Nowadays, clearcoats are widely used in the automotive industry. UV protection must be realized with non-colouring additives capable of absorbing penetrating dangerous radiation and rendering it into a harmless form of energy. UV absorbers are intended to fulfill this task. They have to act in competition with chromophores also contained in the clearcoat. It is a matter of chance whether an incident photon first meets a chromophore or an UV absorber molecule. This means that chromophores and UV absorber are in competition with each other. This competition can only be shifted in favour of the UV absorber if it is present in high concentration. However, the solubility of UV absorbers in resin is usually only of the order of a few per cent. Higher concentrations can yield a sticky coating surface in outdoor use due to a form of separation known as sweating. This can lead to dirty surfaces and loss of an additive that is not exactly cheap.

Clearcoats have two basic tasks. This is best explained with reference to automotive coatings mostly designed as basecoat/clearcoat systems. Today's clearcoats have to give a car the desired appearance, which is most commonly high gloss, and to protect the basecoat while imparting the attractive colour. Furthermore, the clearcoat has to protect not only the basecoat, but also itself against weathering stress. These two functions have to be considered when light stabilisers are incorporated. Figure 3.10 shows the absorbance of three typical samples from a series of 16 tested clearcoats for modern automotive coatings in the unweathered state (thick curves) and after 14 summer weeks' outdoor exposure in Jacksonville, Florida (thin curves). Free films of the clearcoats were used for this study.

Comparison of the three clearcoats shows that sample #16 afforded the best protection of the basecoat against UV radiation, due to the position of the steep absorbance curve. The most effective ageing wavelength range of global radiation, below 350 nm, was completely absorbed. Since the curves, measured before and after weathering, did not reveal any considerable change, the protective function of clearcoat #16 was not influenced by the outdoor exposure. The curves for clearcoat #03 in the same wavelength range show that 5 to 10 % of the incident radiation reached the basecoat. The influence of weathering can also be neglected. Clearcoat #01 afforded the worst protection. Of the most destructive wavelength of global radiation at 320 nm, some 20 % reached the basecoat in the unweathered state and as much as 40 % in the weathered state. Figure 3.10 also shows that the basecoats of today's automotive coatings can be adequately, but not completely, protected against weathering stress by means of appropriate clearcoats in a common film thickness of about 30 μm.

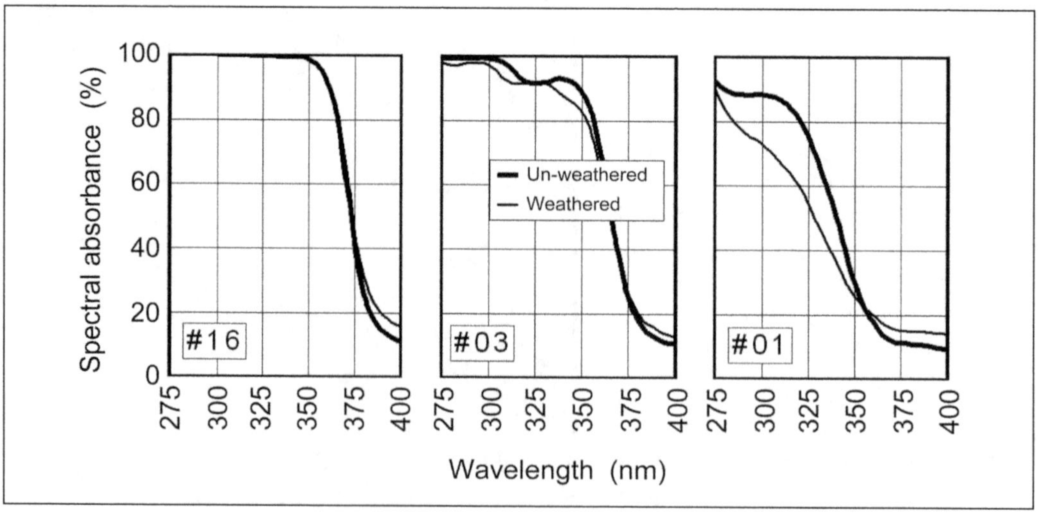

Figure 3.10: Spectral absorbance of films of clearcoat for automotive coatings (#01: Acrylic/melamine, #03 & #16: 2-pack PUR), containing UV absorbers of different effectiveness, measured before and after 14 weeks of outdoor weathering in Jacksonville, Florida

According to the Beer-Lambert law, the radiation incident in a transparent material decreases with increasing film thickness due to radiation absorption. Calculation of the UV radiation as a function of the film thickness for clearcoat #3 using the absorbance value at 320 nm from Figure 3.10 yields the curve shown in Figure 3.11. Some 5 % of the incident radiation reaches the basecoat beneath a 30 nm-thick clearcoat film.

In this context, consider the usual protective coatings designed to be the outermost layers of scratch-resistant polycarbonate car windows; such layers are usually only a few micrometers thick. In this case, protection of polycarbonate, which is known for its moderate weatherability, is more complicated. According to the Beer-Lambert law, such thin surface layers are barely capable of absorbing enough radiation to confer adequate UV protection. Technological reasons preclude both an abnormally high UV absorber concentration and a higher film thickness. Car windows designed in this way may have very good scratch resistance, but barely sufficient weatherability. Weathering stress may lead to under-chalking by degraded polycarbonate and thus premature tarnishing of the substrate or delamination of the coating.

As far as the protective function of UV absorbers on the clearcoat itself is concerned, the key facts are identical with those for pigments. Here, again, the use of UV absorbers confers only weak protection on the outermost micrometers of the clearcoat surface in line with the Beer-Lambert law. Thus, UV photons can penetrate more deeply, meeting chromophores on their way and generating free-radicals. For this reason, UV absorbers are mostly used in combination radical scavengers and antioxidants. These special light stabilisers have the task of neutralising the free-radicals formed by the initiation reaction of photooxidation (see reaction (1) in Figure 3.2) and preventing the propagation of the free-radical chain

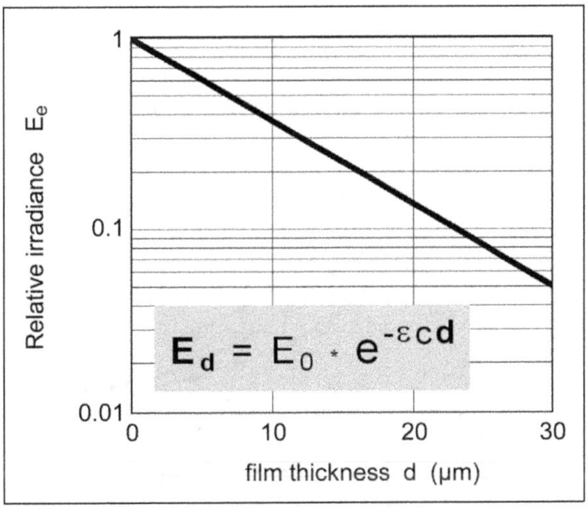

Figure 3.11: Absorption of radiation according to the Beer-Lambert law (ε: extinction coefficient, d: film thickness, E_0: radiation intenstiy at d = 0, E_d: radiation intensty of the transmitted radiation)

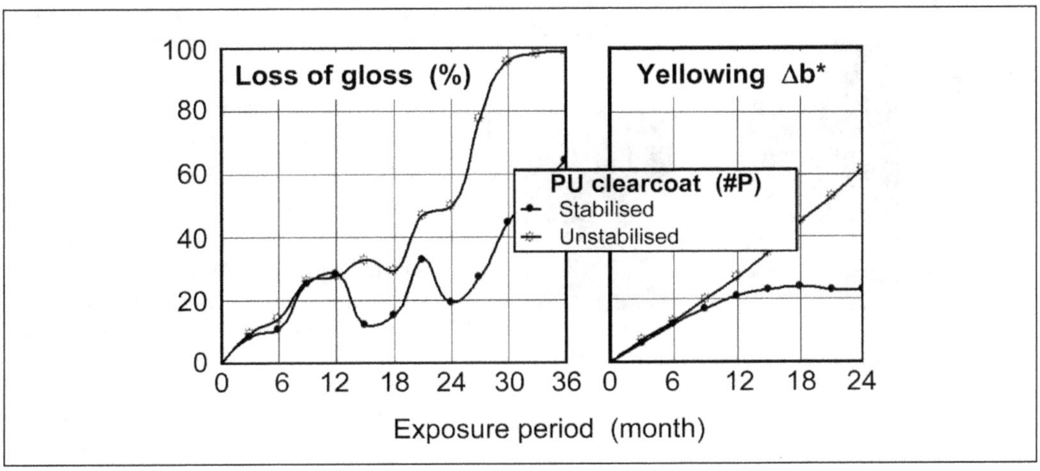

Figure 3.12: Ageing of a PUR-based white basecoat/clearcoat system with and without light stabilisers during conventional outdoor weathering in Miami, Florida (adapted from [22])

mechanism that is also shown in Figure 3.2. In contrast to UV absorbers, which work by physical means, free-radical scavengers catch free-radicals chemically without suffering damage themselves. The best known representatives of this type of light stabiliser in the paint chemistry are sterically-hindered amine light stabilisers (or HALS). Nowadays, adequate protection against the destructive action of UV radiation can only be achieved through a proper combination of both UV absorbers and HALS. Figure 3.12 shows how the presence of UV absorbers and free-radical scavengers in the clearcoat influences the ageing rate of a PUR-based basecoat/clearcoat system during outdoor weathering in Miami, Florida, by comparing the ageing behaviour of stabilised and unstabilised samples.

The diagram on the left shows the efficiency of light stabilisers in respect of loss of gloss. Ageing of the stabilised sample proceeds considerably lower than that of the unstabilised one, revealing the protective effect of the light stabilisers. While the unstabilised sample has lost nearly all its gloss after 36 months of exposure, the stabilised one has retained about 40 % of its initial gloss. Yellowing serves as the ageing criterion in the diagram on the right. Assuming that the clearcoat does not suffer any change in colour, and yellowing occurs only in the white-coloured basecoat, it can be assumed that the continuous increase in yellowing of the unstabilised sample is caused by the almost unhindered passing of the intruding UV photons through the clearcoat due to the absence of UV absorbers. In contrast, yellowing of the stabilised sample ceased after 12 months of exposure. Perhaps not enough UV photons reached the basecoat due to the presence of UV absorbers inside the clearcoat.

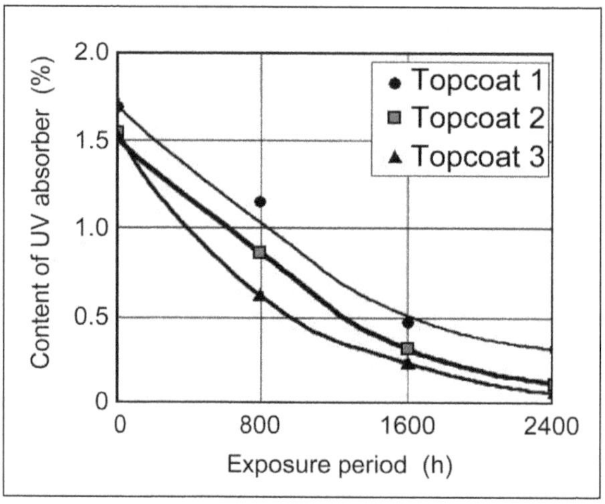

Figure 3.13: Weathering-induced loss of UV absorber in different automotive topcoats (adapted from [23])

These deliberations on the mechanism of UV protection employed in today's coatings should be of some assistance with recognizing sources of uncertainties in accelerated testing that may arise from the presence of light stabilisers. For instance, unrealistic test conditions can cause UV absorbers to evaporate and therefore lead to premature loss of an essential part of the coating's UV protection. Figure 3.13 shows one example of three different automotive coatings whose content of UV absorbers can be completely lost under certain circumstances during accelerated testing.

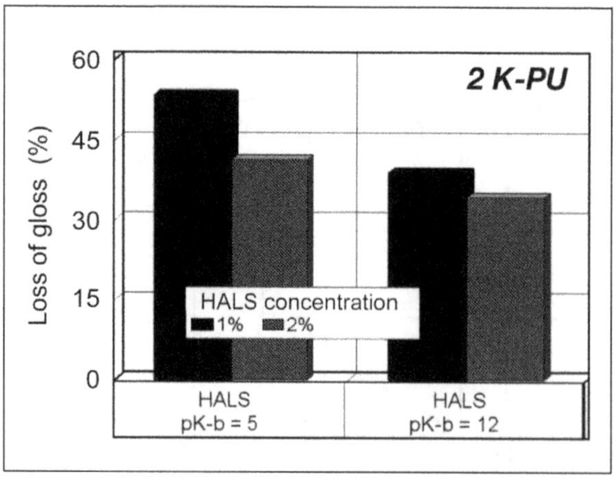

Figure 3.14: Loss of gloss in a 2-pack PUR clearcoat stabilised by HALS of various basicity values (pK-b) and stabiliser concentrations as a result of artificial weathering including acid stress.

Furthermore, with regard to both UV absorbers and HALS, undesirable side reactions may occur with other paint components that could adversely affect the efficiency of the UV protection system. Unrealistically high test temperatures could be one cause of them. HALS can additionally be affected by acid atmospheric precipitation because they are mostly basic in chemical nature.

Figure 3.14 shows the loss of gloss of samples of a 2-pack PUR clearcoat containing HALS as a function of its concentration and basicity, as a result of a special form of accelerated testing, including stress by artificial acid dew and fog (ADF test, see Chapter 6.2.4). Comparison of the columns of samples having identical HALS concentrations shows that in both cases, those coating samples containing the more basic HALS (corresponding to less pK-b) aged at a faster rate. Loss of efficiency or degradation of HALS cannot be excluded under acid stress if types of higher basicity become neutralised. In that event, coating damage is promoted by penetration of an aqueous medium. In order to destroy the stabiliser, the acid has to make contact with the HALS molecules. Diffusion processes are therefore involved in coating damage. In other words, the use of unrealistically high acid concentrations to simulate acid atmospheric precipitation leads to weathering results which do not reflect what actually occurs in practice. The same result would be obtained in special forms of accelerated testing involving aggressive airborne pollutants where the concentration of gas is too high. On the other hand, to exclude these weathering stress factors from accelerated testing might lead to results that fail to correlate with ageing in normal use in industrial and conurbations.

3.1.5 *Influence of the coating's history*

Coatings often fail in normal use, even though they have proved sufficient weatherability or service life in preceding artificial weathering. Rightly or wrongly, **poor transferability** of laboratory weathering results is usually cited as the reason. Poor conformity of weathering results can also be observed if the same accelerated testing is later repeated on identical coatings taken from the same lot. Another frequent observation is that repeating the same accelerated testing on identical coatings from a later lot unexpectedly yields deviating results, too. In other words, it is not impossible for coating samples which are expected to be chemically identical to exhibit major differences in weatherability when simultaneously exposed to the same accelerated testing. If the accelerated testing was carried out in the laboratory, the coating manufacturer would blame the staff involved in coatings testing, suggesting insufficient accuracy while the test was being performed. If the weathering tests being compared were carried out outdoors, non-conformities in the weathering results would be explained, rightly or wrongly, by the **poor repeatability** of natural weathering stress in general. Conversely, the testing laboratory could blame the coating manufacturer, suggesting poor accuracy when the test panels were being produced.

Possible reasons for differences in ageing behaviour can be unnoticed variations or events during production and preparation of the test panels which occurred before the test panels reached the testing laboratory or the outdoor weathering site. Of course, the testing laboratory has no influence over this and such influences are said to be influences arising from the coating's history. The following examples demonstrate this kind of influence on coating ageing. The first example in Figure 3.15 shows how individual fluctuations in viscosity of the wet coatings and the film thickness of the test panels influence the

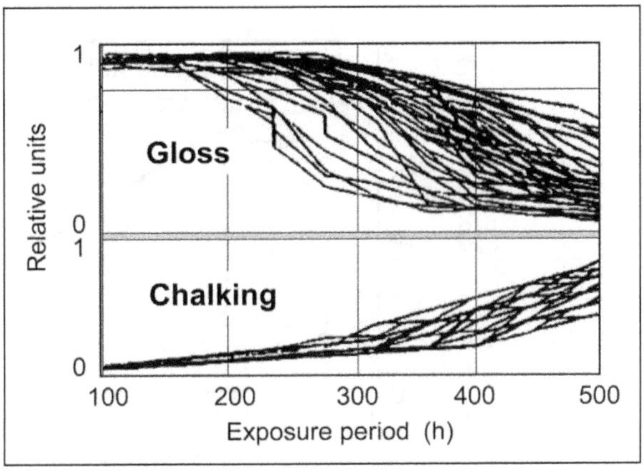

Figure 3.15: History-induced fluctuations in the rate of ageing during accelerated testing of test panels, all painted with the same rutile-pigmented coating (adapted from [24])

weathering results. This figure shows the change in gloss and chalking of numerous specimens made from the same rutile-pigmented alkyd-resin-based coating during artificial weathering. Since the fluctuations in the aforementioned conditions of application were not spotted, accelerated testing revealed great fluctuations in the rate of ageing between the tested specimens.

In this case, a service-life prediction based on 50 % loss of gloss acting as ageing criterion would lead to values for time to failure that varied by as much as 100 %. A further example showing the influence of history comes from the automotive industry. Here, narrow tolerances as regards temperature and duration must not be exceeded during curing. Figure 3.16 shows a case of three different topcoats, in which an increasing curing temperature leads to increasing loss of UV absorbers. In this case, raising the curing temperature from 130 °C to 160 °C would increase the loss of UV absorber by more than 10 %. As discussed in Chapter 3.1.4, loss of light stabiliser can always be expected to lead to accelerated photochemical ageing.

The next example shows that the storage period for prepared test panels before accelerated testing can lead to changes inside the coating material that influence coating ageing.

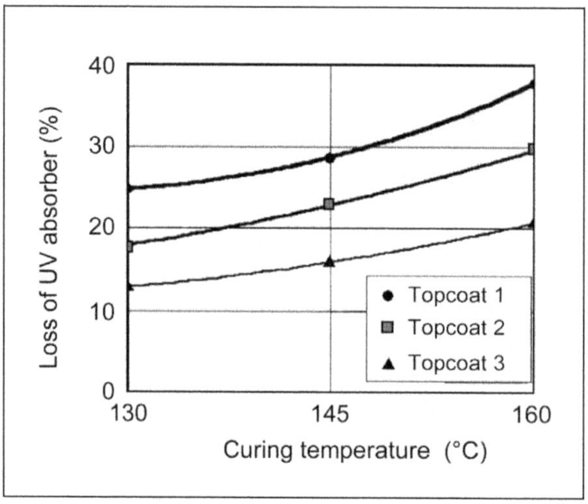

Figure 3.16: Curing-induced loss of UV absorber in automotive topcoats as a function of curing temperature, adapted from [23]

Figure 3.17 utilises the example of an alkyd-resin-based clearcoat to illustrate how the glass transition temperature and the residual solvent concentration change during storage at 100 °C. Both of these coating properties can considerably influence the coating's ageing behaviour. The curves reveal that, in spite of the relatively high storage temperature, the coating needs over one year (10,000 h) to lose its solvent completely and reach its final glass transition temperature. In general, the older the specimen, the better is its weatherability. Oxygen diffusion inside the coating film is hampered with increasing age due to the increasing glass transition temperature.

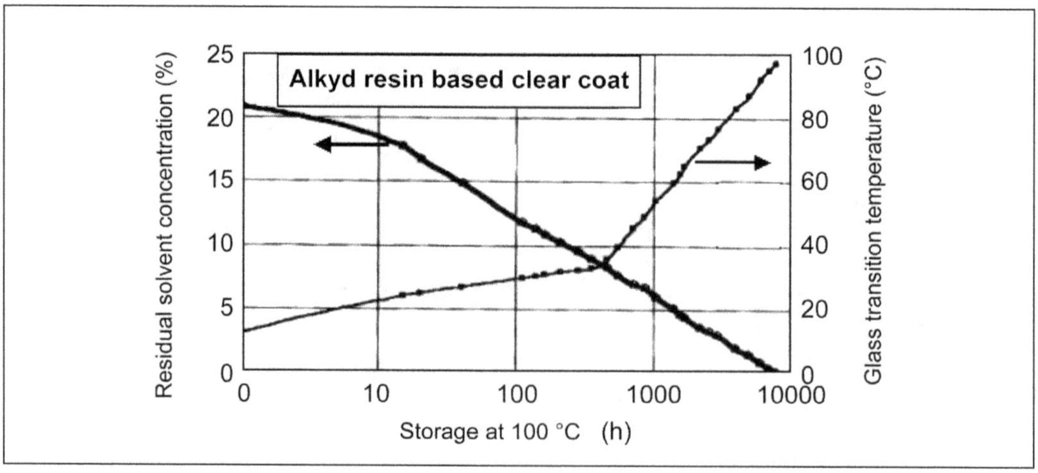

Figure 3.17: Residual solvent concentration and glass transition temperature of an alkyd-resin-based clearcoat as a function of storage at 100 °C (adapted from [25])

3.2 Weathering stresses

3.2.1 Global radiation – main weathering stress

3.2.1.1 Dominant role of global radiation in coating ageing

As already discussed in Chapter 2.1 and shown in Figure 2.1, global radiation can be seen as the dominant weathering stress factor acting on outdoor coatings. As mentioned before, the photons of the global radiation are necessary to provide the energy to activate polymer molecules, to form free-radicals and to trigger photooxidation. In other words, no global radiation means no photochemical ageing.

With this in mind, the properties of global radiation those are required for coating ageing will now be discussed first. Its most important quality features are its spectral distribution and the ratio of its direct to its diffuse portions. Concerning its quantitative properties, only the data related to the irradiated specimen surface are of interest for accelerated testing. Discussion of the qualitative features will therefore be limited to the terms radiation and radiant exposure in this book, as presented in Table 3.5.

Irradiance characterises the strength of the radiation impinging on the surface as expressed by the SI unit W/m² and indicated by the symbol E_e. Radiant exposure is the irradiance integrated over a certain time period, expressed by the SI unit J/m², and indicated by the symbol H_e. Expressing these radiation parameters in relation to a narrow wavelength range of one nanometer yields the spectral irradiance, measured in W/m²/nm, and the spectral radiant exposure, measured in J/m²/nm. The symbols are completed by adding the respective wavelength λ (for instance $E_{e\text{-}\lambda}$ and $H_{e\text{-}\lambda}$)

3.2.1.2 Spectral distribution of the global radiation and activation spectrum

The spectral distribution has proven to be a particular characteristic of the quality of global radiation that requires special attention in the context of accelerated testing.

Table 3.5: Basic terms of irradiation and their SI units found in DIN 5031 [26]

Irradiation parameter	Symbol	SI-unit
Irradiance	E_e	W/m²
Radiant exposure	H_e	J/m²
Spectral Irradiance	$E_{e\text{-}\lambda}$	W/m²/nm
Spectral radiant exposure	$H_{e\text{-}\lambda}$	J/m²/nm

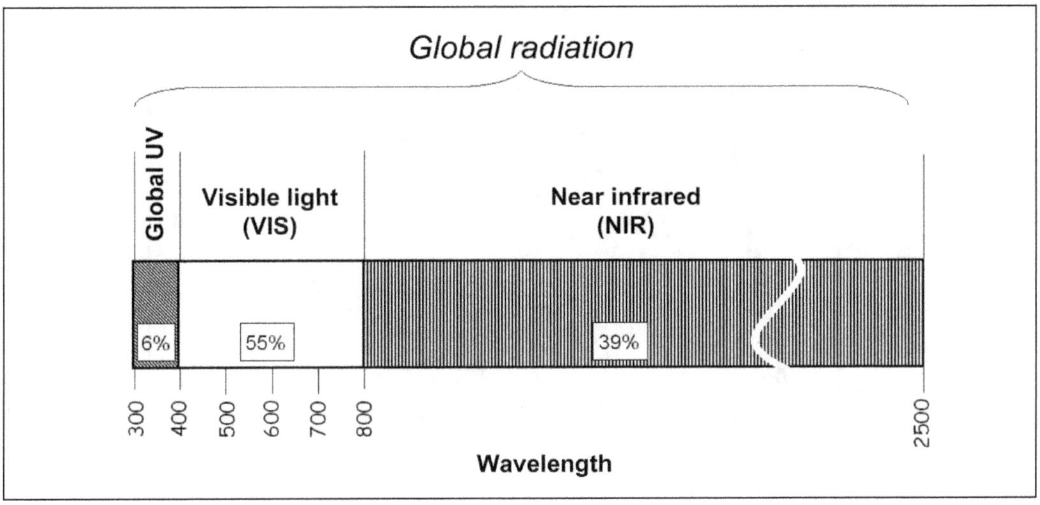

Figure 3.18: Spectral distribution of global radiation

As outlined in Figure 3.18, global radiation can be subdivided into three essential wavelength ranges. These are: ultraviolet radiation from 300 to 400 nm (herein referred to as global UV), visual light from 400 to 800 nm (VIS), and near-infrared from 800 to 2500 nm (NIR). In American literature, the division between global UV and VIS can usually be found at 385 nm. This changes the figures by a factor of 1.3. Frequently, the lower limit of global UV is taken to be 285 nm or 290 nm. Since only very little radiation of wavelength <300 nm reaches the surface of the earth, the differences in irradiation parameters can be neglected in this case. Global UV may also be divided into UVA (from 320 to 400 nm) and UVB (<320 nm).

As already mentioned, global UV (300 to 400 nm) is the most important sub-range of the global radiation where accelerated testing is concerned. Further discussion will therefore concentrate more on global UV than on VIS und NIR. VIS radiation can be perceived by the human eye, whereas NIR is sensed by the skin. In contrast, global UV, which accounts for just 6 % of global radiation, cannot be perceived directly by human beings. Global UV manifests itself only afterwards through its destructive action on organic substances, where it takes the form of sunburn or, in serious cases, skin cancer.

Solar radiation has to penetrate the earth's atmosphere before reaching the earth's surface. It is thus substantially attenuated. Only 80 % max. of extraterrestrial solar radiation (radiation prior to entry into the atmosphere) reaches the earth's surface. On its journey, solar radiation can be influenced by many attenuating factors. For this reason, global radiation can vary in its irradiance, as well as its spectral distribution. To facilitate comparisons, the international organisation CIE has defined a solar spectrum in Table 4 of its CIE 85 guideline [27]. This is shown in Figure 3.19, together with the spectrum of extraterrestrial radiation. The CIE 85 curve represents the maximal irradiance of global radiation that would be experienced by materials on a horizontal surface at the equator close to noon on a clear day at the spring or autumn equinox. In the context of accelerated testing, this idealised spectrum broadly acts as a reference for assessing artificial radiation sources. Concerning the definition and simulation of worst case weather scenarios, as discussed in Chapter 2.5, the CIE 85 curve can serve as the natural extreme of the weathering stress factor global radiation.

The term activation spectrum is frequently encountered in the literature. In the aforementioned contexts, the reader has already been confronted with the phrase photon of adequate energy. This means that not all of the wavelength ranges contribute to coating ageing to the same degree. The spectral distribution has therefore already been defined as an essential qualitative characteristic of global radiation. Photon energy (E_p) is related to the wavelength (λ by Planck's

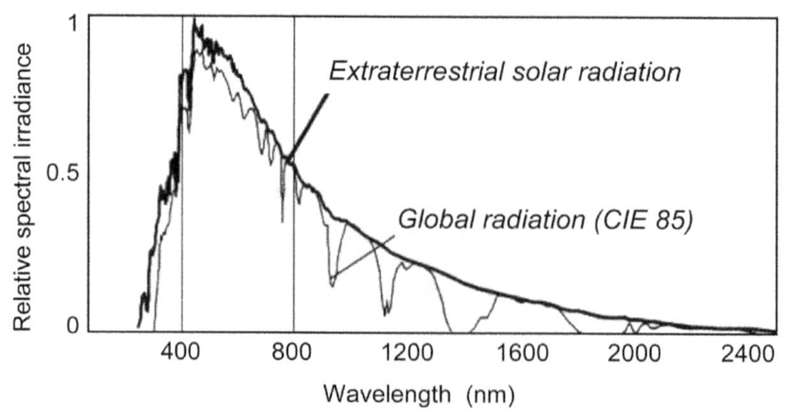

Figure 3.19: Comparison of extraterrestrial and global radiation based on CIE 85, Table 4

law: $E_p = h\ c/\lambda$ (h = Planck's constant, c = speed of light). In other words, photon energy increases with decreasing wavelength. **The shorter the wavelength, the greater is the ability to cleave more stable polymer bonds and increase the ageing rate.** For a given polymer or coating, the ageing rate as a function of wavelength is known as the spectral sensitivity and indicated by the symbol s_y.

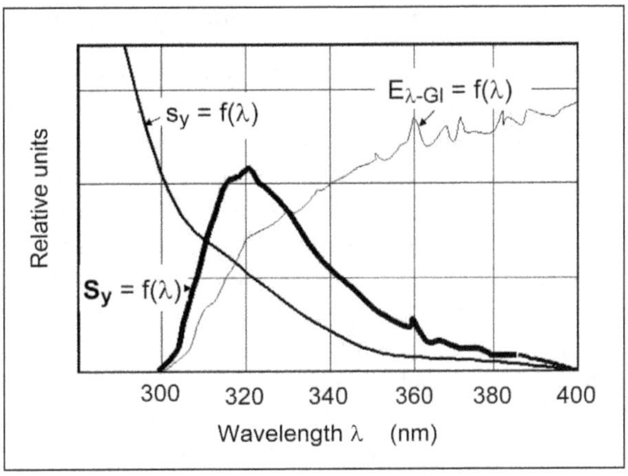

Figure 3.20: Polymer-typical activation spectrum ($S_y = f(\lambda)$) of global radiation resulting from the spectral sensitivity of the polymer ($s_y = f(\lambda)$) and the spectral distribution of the global radiation ($E_{\lambda\text{-}Gl} = f(\lambda)$) for the wavelength range up to 400 nm

The curve of spectral sensitivity as a function of wavelength ($s_y = f(\lambda)$) in Figure 3.20 is typical of polymeric materials and is characterised by the steep slope towards shorter wavelengths. Almost identical curves can be found for most of the common paint resins. In other words, the shorter the wavelength, the more destructive is the radiation. Differences occur with regard to the steepness of the slope, as well as the upper limit of effectiveness as a function of coating type.

Now, let us return to the activation spectrum. It is characterised by $S_y = f(\lambda)$ and is obtained by multiplication of the relative spectral sensitivity $s_y = f(\lambda)$ with the global radiation ($E_{\lambda\text{-}Gl} = f(\lambda)$), for instance, based on CIE 85. It should be remembered that the activation spectrum calculated in this way is valid only for the global radiation resulting from the sun as radiation source, and especially only for a specific type of global radiation, namely that specified in the CIE 85 curve. If the spectral distribution of the radiation used for accelerated testing is altered, for instance, through the use of filtered sunlight or the use of various artificial radiation sources, other activation spectra can result. Furthermore, the activation spectrum must not be seen as an intrinsic property of the radiation source used, but rather as a combined property created by a characteristic of the irradiated material (spectral sensitivity) and a characteristic of the radiation source employed (spectral distribution, also often called the radiation spectrum).

Now for some words on the activation spectra created by global radiation as representing outdoor conditions. In general, they are characterised by a maximum value near 320 nm, due to the strongly decreasing number of photons of wavelengths shorter than 320 nm (reaching zero at 300 nm). The maximum is that wavelength of global radiation which is most destructive of polymers. This fact needs to be given serious consideration in the design of artificial methods of accelerated testing and the choice of radiation source. It should be remembered that the aim of accelerated testing is to shorten the time necessary for obtaining usable test results. This can only prove to be the case if sufficient radiation of 320 nm is delivered by the radiation source. As UV radiation, and especially that portion of wavelength close to 320 nm, contributes most to coating ageing, the subdivision of global radiation shown in Figure 3.21 will be used in the rest of this book. This subdivision is adequate for accelerated testing. The influence exerted by wavelength is reflected by subdivision into **global radiation** (total), comprising the wavelength range from 300 to 2500 nm; **global UV**,, comprising the wavelength range from 300 to 400 nm; and **global UV at 320 nm**, comprising only the narrow wavelength range from (320 ±0.5) nm. Further subdivision to distinguish between direct and diffuse radiation is described in the next paragraph.

3.2.1.3 Direct and diffuse portions of global radiation

As shown in Figure 3.21, global radiation has also been subdivided into its direct and diffuse portions. This subdivision has been performed for all the wavelength ranges depicted here, too. As discussed in Chapter 3.2.1.2, the reference spectrum of global radiation described in Table 4 of CIE 85 is only valid for the standard conditions defined there. These standard conditions characterise a special instance of natural stress by irradiation. If the sun serves as radiation source for accelerated testing, as is the case for all types of outdoor weathering, considerable deviations from this reference spectrum must be reckoned with. Global radiation varies in the ratio of irradiance and spectral distribution. As a result, the rate of coating ageing can differ from locality to locality, hour to hour, day to day, and so on. Furthermore, the ratio of direct to diffuse radiation plays another important role. This ratio is predominantly influenced by the length of the path that the solar radiation traverses from sun to earth. The path through the atmosphere is usually characterised by the term air mass (AM)) in meteorology and serves as a relative value. Its functional relationship to the sun's angle of elevation γ is given by the equation AM = $1/\sin \gamma$. The angle of elevation $\dot{\gamma}$ marks the variation in time and position of the sun above the horizon, as outlined in Figure 3.22.

In this diagram, the increase in the length of the path traversed by the sun's rays (drawn in black here) through the atmosphere (grey) with decreasing angle of elevation is clearly recognizable. The sun's angle of elevation varies with the geographical position, season and time of day. The qualitative properties of global radiation therefore vary in the same way. For instance, global radiation reaches its maximum highest irradiance of about 1000 W/m^2 at noon on the summer solstice when the sun's angle of elevation has reached its maximum value of 66.5 degrees (at 47 degrees northerly latitude). With regard to the whole world, global irradiance reaches its maximum near the Tropic of Cancer in June and near the Tropic of Capricorn in December.

Besides absorption, solar radiation experiences more or less strong scattering on its path through the atmosphere. Due to this effect, some of it is diffuse when it reaches the earth. It is therefore necessary to differentiate between the direct and diffuse portions when characterising global radiation. Direct radiation means that portion which arrives on the earth by the shortest route through the atmosphere. A planar test panel would receive the maximum possible irradiance by direct radiation if it were arranged perpendicularly to the angle of incidence of the sun's rays, i.e. if it directly addressed the sun. Where there are deviations from the ideal angle of incidence,

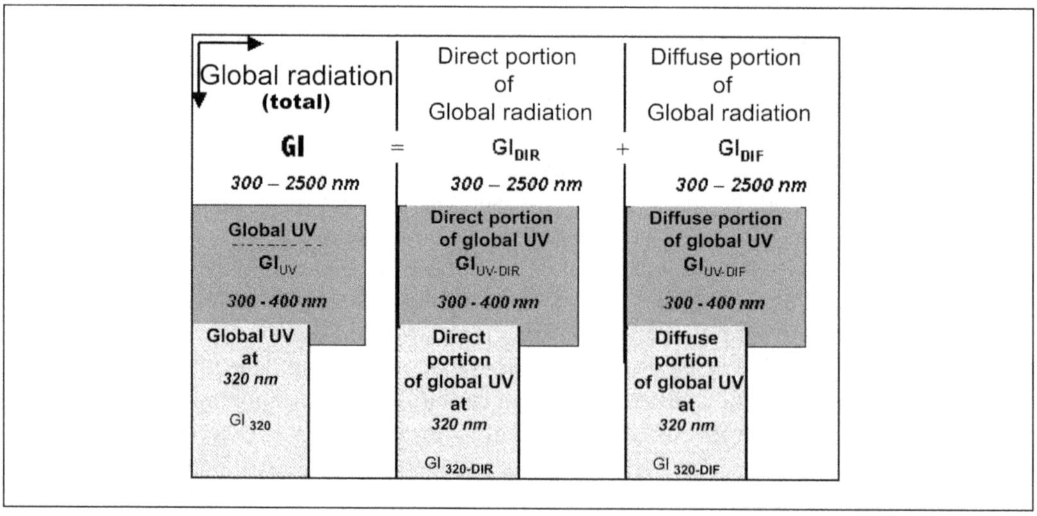

Figure 3.21: Simplified subdivision of global radiation being sufficient for this book's purposed

reduced values of irradiance by direct radiation must be expected according to the Lambert Cosine law.

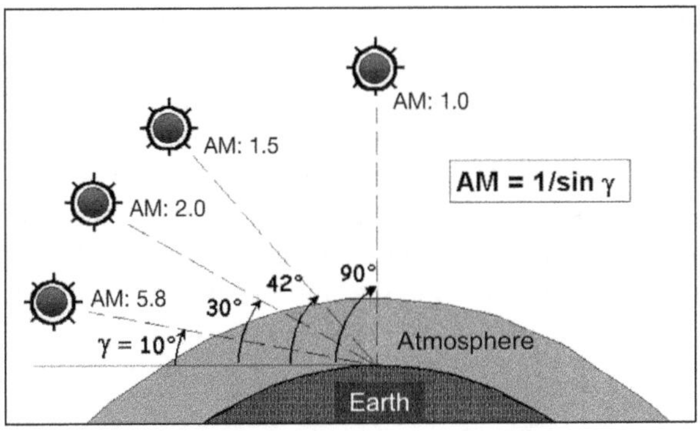

Figure 3.22: Relationship between sun's angle of elevation "γ" and air mass AM traversed passed by solar radiation

The diffuse portion of global radiation, also known as diffuse sky radiation, would be received by test panels arranged horizontally and exposed to the full half-space of the sky. According to Rayleigh's law, scattering is proportional to λ^{-4}. This means that shorter wavelengths are preferentially scattered and longer wavelengths are preferentially transmitted. Therefore, scattering is the predominant effect on the global UV most responsible for coating ageing. This explains why you can suffer from sunburn if you spend your time in a shady spot protected against the sun rays. A shady spot on a sunny day provides poor shelter against the highly scattered UV radiation emanating invisibly from all directions of the full half-space of the sky. The other example concerns the VIS wavelength range. The blue colour of the sky is a result of scattering. Its blue colour on a sunny day can be explained by predominant scattering of blue light due to its shorter wavelength.

In Figure 3.23, spectra for the direct portion of global radiation are presented as a function of the sun's angles of elevation in the wavelength range from 300 to 800 nm (UV and VIS). The curves were calculated by means of the CESORA Model Smart 2 program, commercially available software described in [28]. It enables the portions of direct and diffuse irradiance to be readily calculated for each exposure angle, each geographical location, each degree of cloud cover, and each point in time. As can be seen in Figure 3.23, the three spectra differ enormously. The extreme decrease in direct radiation in the range of global UV clearly reveals the influence of the wavelength. With regard to the VIS wavelength range, the curves reveal a shifting of the spectra towards red with decreasing angle of elevation of the sun.

In Figure 3.24, direct and diffuse portions are shown as a function of the sun's angle of elevation for the two narrow wavelengths ranges (400 ± 0.5) nm and (320 ± 0.5) nm. As can be seen, the direct portion decreases strongly and the diffuse portion increases strongly as the angle of

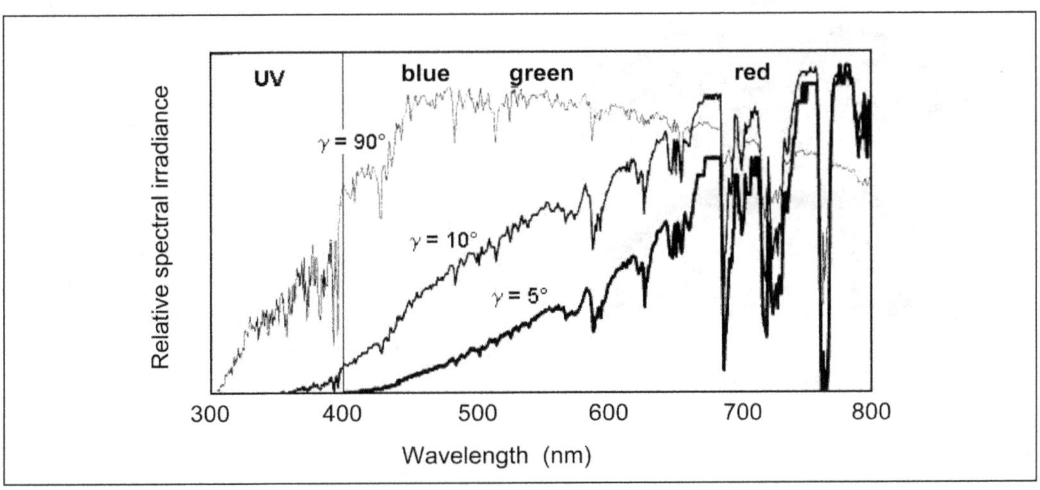

Figure 3.23: Relative spectral distribution of the direct portion of global radiation in the wavelength range up to 800 nm for the sun's angles of elevation γ=5°, 10°, and 90°

elevation decreases. Comparison of the two diagrams reveals that the shorter-wavelength radiation undergoes much more scattering, as evident from the lower ratio of direct to diffuse radiation. For a sunny summer's day in Central Europe (for instance, at 42° north, AM = 1.5), it can be inferred that direct radiation preponderates at 400 nm, but diffuse radiation preponderates at 320 nm. With respect to accelerated testing, these considerations are particularly important where the sun serves as radiation source. They should help in the assessment of weathering stress arising during the different types of outdoor weathering. Ignoring scattering can lead to problems when the weathering results are transferred to the intended normal use.

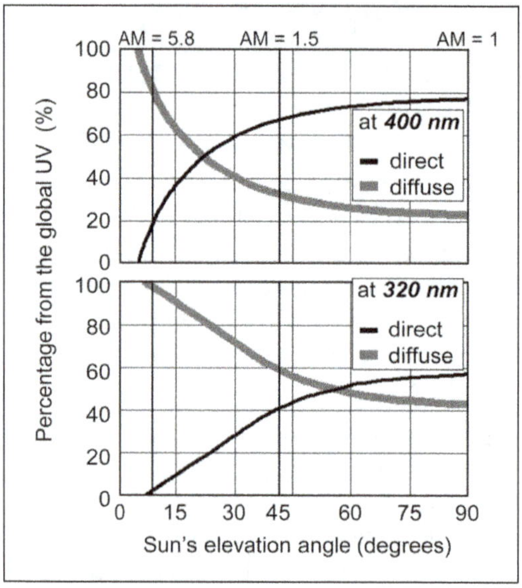

Figure 3.24: Direct and diffuse portions as a function of the sun's angle of elevation and the wavelength as illustrated by two narrow ranges of global UV (400 ± 0.5) nm and (320 ± 0.5) nm (calculated by means of CESORA)

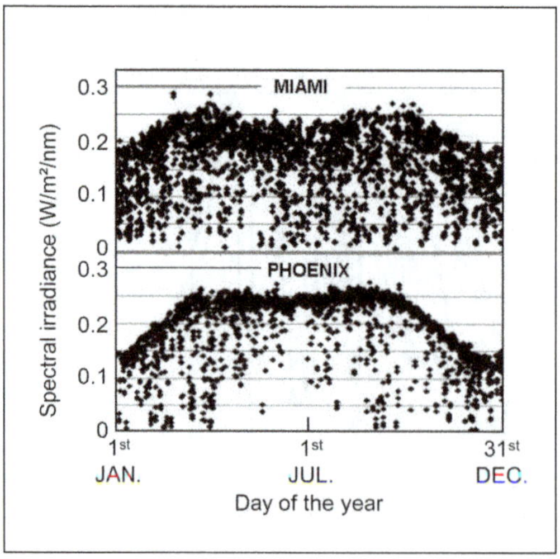

Figure 3.25: Spectral irradiance at 316 nm plotted at 45° South in Miami, Florida, and Phoenix, Arizona, at noon from 1997 to 1999 as a function of the day of the year (adapted from [29])

3.2.1.4 *Influence of clouds on global radiation*

Clouds are an essential barrier to the passage of solar radiation through the atmosphere. This is readily apparent from the strong decrease in brightness when a cloud passes in front of the sun. Aside from the air molecules, the water molecules absorb and scatter the radiation. Where the sun serves as the radiation source for accelerated testing as in the case of all types of outdoor weathering, global radiation will also vary with the degree of cloudiness. **Clouds are the main reason for the poor repeatability of natural weathering stress in general.**

Clouds vary randomly. Widely differing weathering results are therefore the norm. Only the use of mostly unacceptably long exposure periods can be expected to produce somewhat repeatable weathering stress. It is therefore very important that the weathering site for outdoor weathering be chosen carefully. Global radiation at the chosen weathering site should not vary unacceptably, at least, not from year to year. Further considerations include providing the staff involved in coating and testing with the necessary know-how for deciding on the weathering site and the type of outdoor weathering most suited to the purpose. Figure 3.25 shows the extent of fluctuation in global radiation for the North American reference climates, Miami, Florida, and Phoenix, Arizona, which are described in more detail in Chapter 4.2.1. The yearly change in spectral irradiance is shown for a 1 nm wavelength range close to 320 nm as measured at noon every day from 1997 to 1999. It should be remembered that the wavelength of the global UV chosen here is known to be the most destructive to polymers undergoing outdoor weathering! The upper picture shows the results for the subtropical region of Miami, Florida.

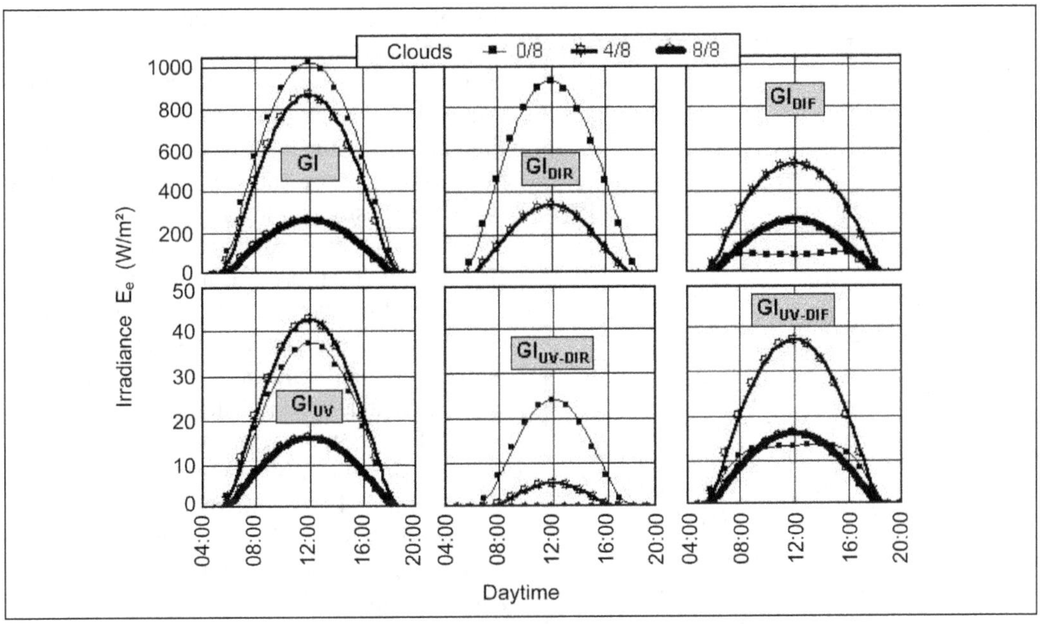

Figure 3.26: Global radiation (Gl) with its direct portion (Gl_{DIR}) and diffuse portion (GL_{DIF}) (top row); and global UV (Gl_{UV}) with its direct portion ($Gl_{UV\text{-}DIR}$) and diffuse portion ($Gl_{UV\text{-}DIF}$) (bottom row), all calculated by means of the CESORA program, for horizontally exposed test panels in Miami, Florida, on 30th June, as a function of the degree of cloud cover and time of day

The humid subtropical climate of Miami, Florida, is characterised by mostly high, very changeable cloud cover. For this reason, the irradiance values, which were always measured at noon, vary between zero and their maximum at about 0.3 W/m²/nm. In this case, the fluctuation in the values seems to be almost independent of the season.

The lower picture of Figure 3.25 shows the results for Phoenix, Arizona, which has an arid desert climate. Since clouds occur less frequently, the fluctuation in irradiance appears to be lower than that for Miami, as witnessed by the denser distribution of the values near the maximum. In contrast to Miami, the fluctuation here appears to be dependent on the season. As expected, the summer months exhibit the least fluctuation in irradiance, while the winter months exhibit similar behaviour at both weathering sites. As we will see below, the ratio of direct to diffuse radiation is also influenced by clouds. The result of corresponding calculations is shown in Figure 3.26 for the example of the reference climate in Miami, Florida.

It shows the influence of the clouds during the day. In line with meteorological practice, the degree of cloud cover is subdivided into three: 0/8 (clear sky), 4/8 (50% covered sky) and 8/8 (completely covered sky). The irradiance varies as a function of the time of day and the sun's angle of elevation. The diagrams in the top row show the total global radiation, Gl (300 to 2500 nm), along with the direct portion, Gl_{DIR}, and the diffuse portion, Gl_{DIF} (see also Figure 3.21). The diagrams in the bottom row show only the global UV Gl_{UV} (300 to 400 nm), again with the direct portion, $Gl_{UV\text{-}DIR}$, and the diffuse portion, $Gl_{UV\text{-}DIF}$. As expected, the global radiation decreases with increase in cloud cover. About one quarter of the irradiance calculated under clear skies can be found in a completely covered sky. Astonishingly, the irradiance decreases by only 20% if the degree of cloud cover increases from 0/8 to 4/8. The second and third diagrams in the top row allow comparison of the direct portion, Gl_{DIR}, and the diffuse portion, Gl_{DIF}, of global radiation. As expected, the direct portion, Gl_{DIR}, attains maximum irradiance under a clear sky and zero irradiance under a completely covered sky. The diffuse portion, Gl_{DIF}, arriving at the earth through the clouds and their gaps, attains maximum irradiance in a half-covered sky (4/8), the value being

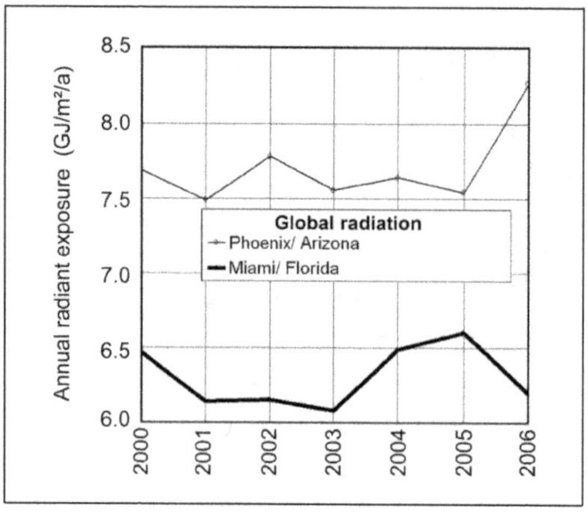

Figure 3.27: Annual radiant exposure (global radiation) measured at 5 ° south, in Miami, Florida, and Phoenix, Arizona, from 2000 to 2006 (adapted from [30])

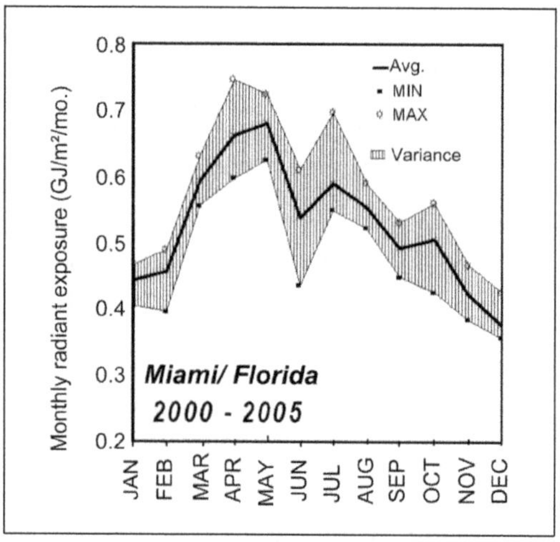

Figure 3.28: Average (Avg.), minimum (MIN) and maximum (MAX) values of monthly radiant exposure (global radiation) measured in Miami, Florida, from 2000 to 2005 (adapted from [30])

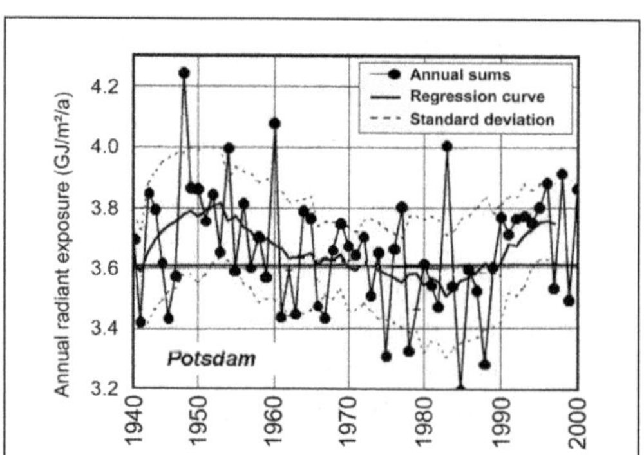

Figure 3.29: Annual radiant exposure (global radiation) measured in Potsdam, Germany, from 1940 to 2000 (adapted from [31])

50 % of that calculated for total global radiation under a clear sky (0/8). Results for global UV (300 to 400 nm) in the bottom row are more interesting. Surprisingly, the global UV, Gl_{UV}, as the sum of its direct, $Gl_{UV\text{-}DIR}$, and diffuse portions, $Gl_{UV\text{-}DIF}$, attains higher irradiance under a half-covered sky (4/8) than under a clear sky (0/8). As expected, comparison of $Gl_{UV\text{-}DIR}$ and $Gl_{UV\text{-}DIF}$ shows that $Gl_{UV\text{-}DIR}$ is more preponderant than the diffuse portion of $Gl_{UV\text{-}DIF}$ under a clear sky (0/8). However, the diffuse portion, $Gl_{UV\text{-}DIF}$, is more preponderant than the direct portion, $Gl_{UV\text{-}DIR}$, by about factor 4 under a half-covered sky (4/8). Astonishingly, under a completely covered sky, the diffuse portion, $Gl_{UV\text{-}DIF}$, still attains almost 40 % of the irradiance calculated for the whole global UV, Gl_{UV}, under the most favourable conditions. Figure 3.27 shows the annual natural radiant exposure measured at the weathering sites Miami, Florida, and Phoenix, Arizona, from 2000 to 2006. As can be seen, it varies by at most 10 %, mostly due to fluctuations in cloud cover. This range of fluctuation is unavoidable and must be expected in outdoor weathering.

Figure 3.28 shows the monthly fluctuation in radiant exposure measured in Miami, Florida. The respective maximum (MAX), average (Avg.), and minimum values (MIN) measured over 6 years are shown. The shaded area indicates the resulting fluctuation. Looking at June, by way of example, monthly radiant exposure varies from about 0.4 GJ/m²/month for the cloudiest June and about 0.6 GJ/m²/month for the sunniest June (about 30 %) over the 6 years. Due to the extent of this monthly fluctuation, limiting a conventional outdoor weathering test to a few months should be rejected as unacceptable.

Now, let us turn our attention to clouds and global radiation in Central Europe. By analogy with Figures 3.27 to 3.30 show data published by a typical German weather research institution.

Figure 3.29 shows the annual radiant exposure measured in Potsdam, Germany,, from 1940 to 2000. Over these 60 years, it has fluctuated by about 25 %, from 3.2 GJ/m²/month (1984) to 4.3 GJ/m²/month (1947). In this case, the fluctuation clearly exceeds the respective values for Miami, Florida. The weather in Potsdam can be seen as typical of regions in central Europe that have no contact with the sea. The next example, Hamburg, Germany, is representative of coastal areas in Central Europe. The data available here stem from measurements covering 40 years.

By analogy with Figure 3.28, Figure 3.30 shows the monthly values of global radiation for the period 1949 to 1988. Again, the shaded area shows the fluctuation from the respective minimum value (MIN) to the respective maximum value (MAX). Looking at the values for June, the monthly radiant exposure varies by a factor of 10 from about 0.1 GJ/m²/month for the cloudiest June and about 1 GJ/m²/month for the sunniest June over the course of 40 years. In comparison to these high values of fluctuation, those found in Miami, Florida, or Phoenix, Arizona, can be seen as insignificant.

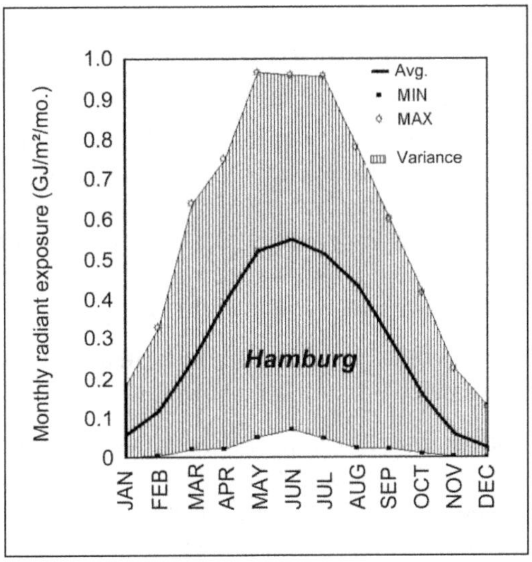

Figure 3.30: Average (Avg.), minimum (MIN) and maximum (MAX) values of monthly radiant exposure (global radiation) measured in Hamburg, Germany from 1949 to 1988 (adapted from 31)

In other words, due to the high, random fluctuation in global radiation in areas of moderate climate and to the random fluctuation in cloud cover, outdoor weathering in Central Europe will always lead to poor repeatability, even if performed over several years. Since higher degrees of cloud cover mostly occur in combination with lower temperatures acting as a further major weathering stress factor, conventional outdoor weathering in Central Europe cannot be recommended for coatings testing.

3.2.1.5 Influence of altitude on global radiation

Since the air mass (AM) traversed by the sun's rays decreases with increasing altitude (elevation above mean sea level, AMSL), lower absorption and scattering of the solar radiation might be expected. However, efforts to use weathering sites in alpine locations to age coatings faster on account of the higher UV irradiance have failed. For instance, a weathering site run on

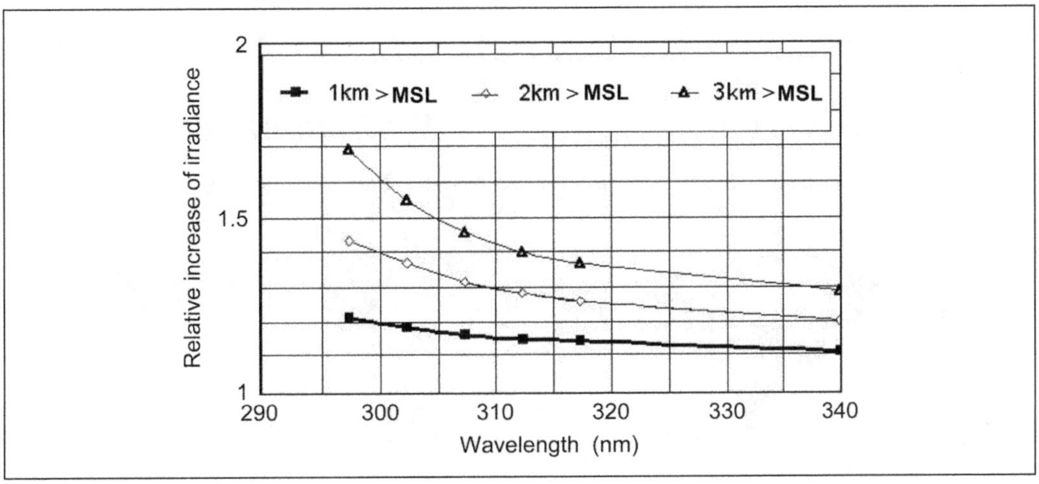

Figure 3.31: Irradiance of global UV as a function of the altitude above mean sea level (AMSL) for the wavelength range up to 340 nm (factor 1 corresponds to MSL) (from [32])

Jungfraujoch, Switzerland (3576 m above mean sea level), yielded UV irradiance values up to 0.93 W/m²/nm at 340 nm which were higher than those measured in Phoenix, Arizona (0.83 W/m²/nm at 340 nm) [12].

However, anybody expecting to obtain a higher ageing rate for accelerated testing at high altitudes will be disappointed. The influence of altitude on UV irradiance is shown in Figure 3.31 for the wavelength range up to 340 nm. Looking at the situation at 320 nm, we see that the irradiance increases by factor 1.35 if the altitude increases from 0 to 3000 m above mean sea level (MSL). However, the weathering stress factor heat does not come into play because temperatures decrease with increase in altitude.

Weathering sites that were built in alpine regions in expectation of a time-saving effect due to greater UV irradiance have failed to acquire any significance in coatings testing.

Tests on aircraft coatings have yielded the same result [21]. As in a weathering study, test panels were fixed on long-distance aircraft and exposed uninterruptedly to normal aviation use. Simultaneously, identical test panels were exposed to conventional outdoor weathering on the ground. Comparison of the weathering results after six years failed to reveal any differences in ageing rate between the two weathering sites. Despite higher UV irradiance at the usual flight altitudes of 10,000 m, the expected time-saving effect did not materialize. Obviously, the low minus temperatures on the plane during the flight seemed to freeze the mechanism of photooxidation (as shown in Figure 3.2), especially, the chemical reactions of polymer free-radicals formed by UV photons. Only that third of the aircraft's life which is actually spent on the ground at airports proved to be responsible for the ageing rate of aircraft coatings.

3.2.1.6 Measurement of global radiation

Anybody who commissions outdoor weathering tests wants to get as much information about the weathering stress as possible. As mentioned before, considerable fluctuations in weathering parameters, especially as regards irradiance date from year to year have been observed. Nonetheless, it is essential to obtain this information. Total global radiation is generally measured by means of pyranometers. Conventional instruments of this type (e.g. complying with ISO 877 [1]) determine the irradiance by measuring the temperature difference between a black and a white receiver surface. The influence of other outdoor conditions, such as rain and wind, is excluded by means of a quartz dome. Operated in the horizontal position, the pyranometer measures the total direct and the total diffuse portions of total global radiation. It is useful to measure the global radiation not only horizontally, but also at other tilt angles. Commonly, the pyranometer is tilted to the exposure angles that customers tend to want. The left picture in Figure 3.32 shows a typical commercial version.

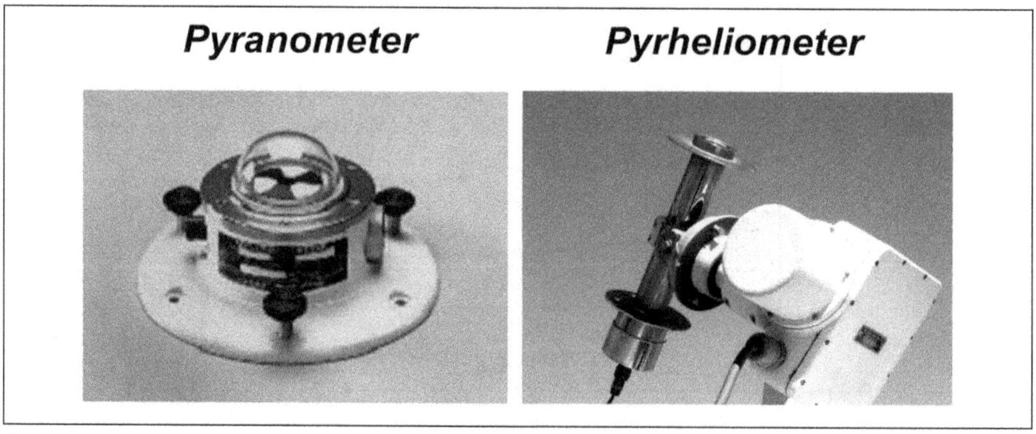

Figure 3.32: Standardised instruments commonly used for measuring radiation during outdoor weathering

Global UV is measured with a UV radiometer. Optical narrow bandpass filters limit the radiation to a specified range of the global UV desired or to higher or lower bandwidths. On account of the greatly reduced wavelength range (<6 % of the total global radiation), highly sensitive photocells are needed to measure the radiation. In order that the ratio of direct and diffuse radiation may be determined, it is usual to additionally measure the direct portion of global radiation. A pyrheliometer, shown in the right of Figure 3.32, is used for this. It is equipped with a precision aperture which only allows passage of a direct beam. The direct portion of global UV is measured in a similar manner.

3.2.2 Thermal stress during weathering (constant/cyclical stress)

As will be described later in Chapter 3.2.5, it makes little sense in the context of accelerated testing to discuss thermal stress in isolation from the other weathering stress factors. In normal use, global radiation, heat, water, and oxygen, which are the most important weathering stress factors, first make contact with the polymer on the coating surface. The temperature on the surface directly influences the rate of coating ageing, and is determined by irradiance, ambient temperature and wind velocity in normal use, as well as most methods of artificial weathering. Separate examination of the influence of temperature is possible only with those artificial weathering methods which suppress the influence of irradiance on the coating surface, for instance, by using fluorescent UV lamps as the radiation source, and which can regulate the relative humidity individually.

Nevertheless, it is worth considering the influence of temperature separately, as it is helpful to staff involved in coatings testing who are seeking the causes of poor repeatability and reproducibility of weathering results. For this reason, the general influence of temperature on photooxidation is described in more detail below. As already discussed in Chapter 3.1.3, the rate of each chemical reaction is predominantly determined by its reaction rate coefficient and the concentration of the reaction partners at the point where the reaction takes place. Again as mentioned before with regard to complex chemical reactions, which include photooxidation it is also a law of reaction kinetics that the slowest step always determines the reaction rate. In simple terms, in the case of photooxidation, the slowest step can be any one of the chemical reactions shown in Figure 3.2 or the coming together of the reaction partners - the polymer free-radicals and oxygen in this instance. First, let us consider the chemical reactions which are influenced by the temperature. In accordance with Arrhenius' law, every chemical reaction comes to a stop if the temperature is low enough (Figure 3.33). According to this law, the reaction rate coefficient, k, increases exponentially with increasing temperature. It is generally found for polymers,

$$k = a \cdot e^{-\frac{E_{act}}{RT}}$$

Figure 3.33: Arrhenius' Law (k: rate coefficient of the chemical reaction, a: temperature- independent constant, E_{act}: activation energy, R: gas constant, T: absolute temperature)

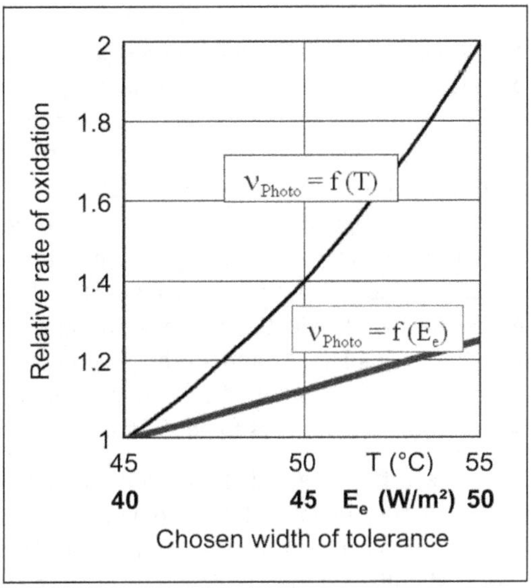

Figure 3.34: Rate of photooxidation (v_{Photo}) as a function of temperature (T) and UV irradiance (E_e) within a chosen tolerance frequently used for artificial weathering performed in simple weathering devices

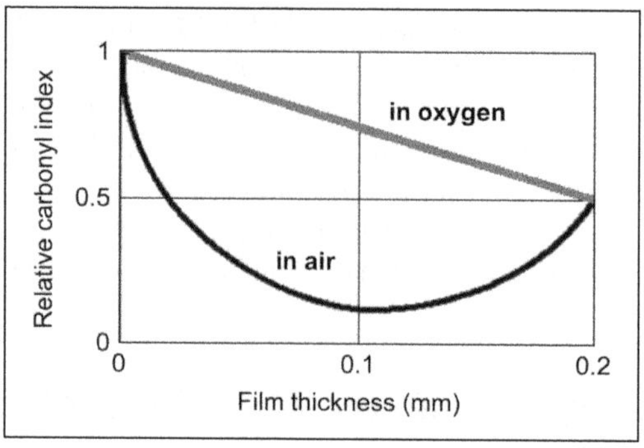

Figure 3.35: Typical curves of the carbonyl index in a polymer film as a function of film thickness during weathering in pure oxygen and air

which include coatings that the reaction rate coefficient will roughly double for every 10 °C increase in temperature or increase by 7 % for every 1 °C increase.

Figure 3.34 shows that fluctuations in temperature have a much greater influence on the ageing rate than equivalent fluctuations in irradiance. The oxidation rate is plotted as a function of temperature and UV irradiance. In this example, narrow temperature and irradiation ranges are chosen to allow for permissible deviations from specified test parameters. These deviations are sometimes used in artificial weathering. As can be seen, coating ageing can increase by a factor of 1.25 within the irradiation tolerance of 45 ± 5 W/m^2. Experience shows that the ageing rate increases almost linearly inside such a narrow range of 10 W/m^2. However, within the temperature tolerance of 50 ± 5 °C, ageing can increase by a factor of 2 on the basis of Arrhenius' law.

There are many examples which indicate that transport of the reaction partners especially that of oxygen penetrating into the polymer, is the slowest step of photochemical ageing. **However, if oxygen is not present at the place where the free-radicals were formed, no photooxidation can take place.**

Oxygen transport into the polymer follows the laws of diffusion. The rate of diffusion depends roughly on the oxygen concentration, c_{O2}, in the ambient atmosphere and on the coeffiecient of diffusion, D. Numerous studies of transparent polymer films of 0.15 to 0.30 mm thickness have demonstrated that the oxygen diffusion is the rate-determining step. Certain coatings can also be found in this range of film thickness. In some cases, polymer films were irradiated in air as well as in pure oxygen. After irradiation, the carbonyl index was measured as a function of the film thickness. Figure 3.35 shows typical curves. The grey curve represents the values after irradiation in pure oxygen. It directly follows the decrease in irradiance with increasing penetration depth in accordance with the Beer-Lambert law already shown in Figure 3.11. In this case, an influence exerted by oxygen diffusion on the ageing rate could not be found. Owing to the high oxygen concentration in the ambient gaseous phase, the rate of oxygen diffusion is high enough, and proves not to be the rate-determining step. At every site inside the polymer where polymer free-radicals are formed, sufficient oxygen molecules are present. In contrast, the black curve for irradiation in air reveals a clear influence exerted by oxygen diffusion. Its curve passes through a minimum of the carbonyl index that always occurs in the middle of the film. Enough oxygen molecules are present only at the front and back sides of the polymer film. There is a lack of oxygen molecules inside the film. Obviously, a considerable number of polymer free-radicals fail to make contact with oxygen molecules and so they recombine with each other, without damaging the polymer. In comparison to irradiation in pure oxygen, the oxygen concentration in air is 21 %. In this case, oxygen diffusion appears to be the rate-determining step.

Therefore, diffusion needs to be included in any consideration of the temperature-dependence of photooxidation. Reliable data for oxygen diffusion are hard to find. Due to the similarity of oxygen and water molecules, data on water vapour permeability could be used instead of oxygen for the sake of simplicity. Data published in [34] on the temperature-dependence of water vapour diffusion in coatings are shown in Figure 3.36.

The curve shows that the water vapour diffusion follows an exponential function with increasing temperature. Surprisingly, the diffusion coefficient, D, stands in the same functional relationship

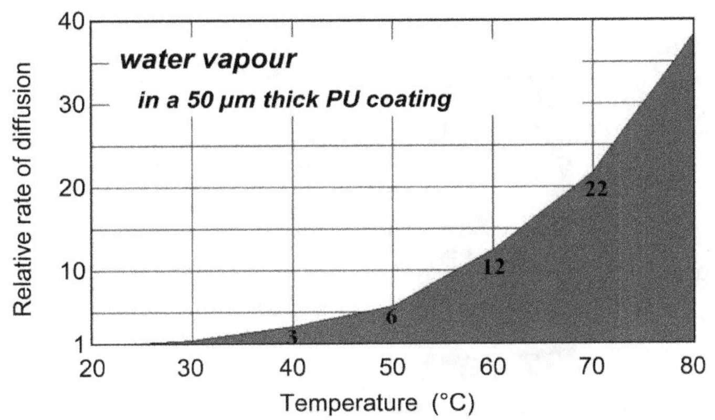

Figure 3.36: Rate of water vapour diffusion in a PUR coating as a function of temperature (adapted from [34])

to the temperature as the reaction rate coefficient, k, namely it doubles for every increase of 10 °C. This means that the functional relationship between the rate of photooxidation and the temperature is independent of whether chemical reactions or diffusion are the slowest step.

Since this relationship is equally valid for water vapour and oxygen, temperature also influences ageing via the stress factor water. As will be described in the following discussions, water catalyses photooxidation by forming special oxygen-containing free-radicals during weathering that are more reactive than the relatively inert atmospheric oxygen molecules. In view of the aforementioned discussions, it does not matter which of the two steps determines the temperature-dependence of the photooxidation. **With regard to accelerated testing, it is essential to repeat the fact that temperature fluctuations can lead to higher changes in the rate of coating ageing than equivalent fluctuations in irradiation.**

Unfortunately, the relationship between ageing rate and temperature, which is based on Arrhenius' law, frequently manifests itself only as an increase in the oxidation products of photooxidation. Important coating properties often do not change in direct proportion to the increase in oxidation products. Furthermore, they often seem to only become visible after a certain concentration of oxidation products has formed. Consequently, Arrhenius' law cannot be used without further ado to calculate the change in certain ageing criteria that would be expected from higher temperatures. This poses the question as to whether raising the testing temperature to a value above the extreme values found in normal use is a proper means of shortening a weathering test. The answer could be "yes" if certain previously proven limits are not exceeded. An overly high temperature could set ageing processes in motion that would never occur in normal use. Temperature-related changes in structure, modification and state due to crosslinking or crystallisation, transgression of the glass transition temperature, relaxation, or even melting can lead to unrealistic changes in ageing behaviour.

Processes caused by unnaturally exceeded temperatures often lead to a product's no longer being identical with the coating to be tested. Unnaturally exceeded temperatures can also cause unnaturally fast depletion of important additives, such as pigments and light stabilisers, by destruction or evaporation, and that can lead to alterations which are never observed in practice. Modern coatings, like automotive coatings, can almost be regarded as similar to living organic matter, because of their high complexity. In order to demonstrate the risk that can arise from an unnaturally exceeded temperature, let us consider an actual example of organic matter, namely a chicken egg. From Figure 3.37, it can be seen that the egg needs 21 days' exposure at 38 °C to form a chicken. According to Arrhenius' law, a temperature of 120 °C would be necessary to shorten the incubating time to one hour. However, the diagram reveals another outcome. This example illustrates the lack of success of acceleration induced by an unnaturally exceeded temperature.

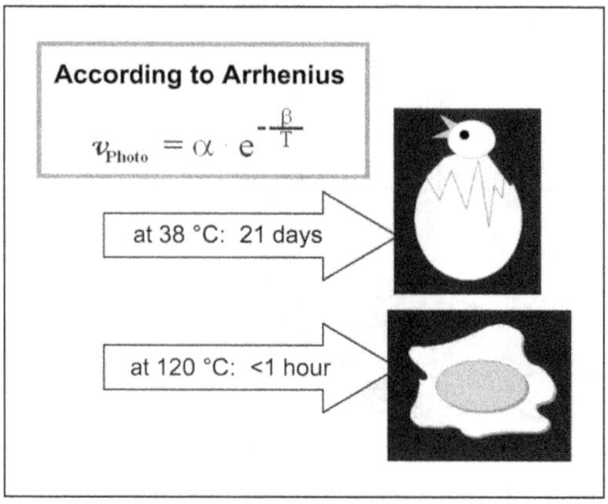

Figure 3.37: Example of unsuccessful acceleration of an organic-chemical process (the chicken egg)

Abrupt changes in the test panel temperature are often seen as a further reason for accelerating coating ageing. The changes are considered to be the result of cyclical mechanical stress caused by cyclical temperature changes. However, it is hard to explain how mechanical stress could influence the free-radical chain mechanism of photooxidation shown in Figure 3.2. In the aforementioned studies on aircraft coatings, test panels were subjected during artificial weathering to abrupt changes in surface temperature from 50 °C to -60 °C several times a day in order that the environmental stress of take-off and landing could be simulated. When the special weathering test was concluded after six weeks, no influence exerted by the additional mechanical stress on the coatings' ageing rate could be proved in comparison to an otherwise equivalent weathering test, which excluded the temperature shocks. Furthermore, performing a higher number of temperature shocks before weathering failed to reveal an influence exerted on the coatings' ageing.

Mechanical strength is an important ageing criterion for plastics such as agricultural film. Coatings also suffer a loss of mechanical strength due to ageing by weathering. With regard to automotive coatings, weathering-induced loss of mechanical strength manifests itself, for instance, in decreased scratch resistance.

However, there is no reliable evidence to the effect that coating ageing is accelerated by cyclical thermal stress. Only when the mechanical strength has been diminished by weathering to such an extent that the polymer can no longer resist the cyclical mechanical stress do temperature shocks reveal hidden ageing effects, for instance, in the form of cracks. In other words, additional mechanical stress does not accelerate photochemical ageing. Loss of mechanical strength due to progressive weathering is merely recognized earlier. Therefore, the inclusion of temperature shocks in accelerated testing makes sense only if mechanical strength (for instance, the time to elapse until cracks form) is chosen as an ageing criterion. In that event, regular temperature shocks could serve as a tool for shortening accelerated testing - not because they accelerate coating ageing, but because they make the ageing visible earlier.

3.2.3 Weathering stress factor water (constant/cyclical effect)

As already described in Chapter 3.1.3, the presence of water may be necessary for some ageing reactions to occur, depending on the polymer to be tested. For most coatings, water is an extremely important weathering stress factor. It occurs in the form of relative humidity in which it is dissolved in air, and in the form of aqueous precipitation, such as rain, dew and fog. The stress factor water is often referred to as moisture. Water dissolved in air is best characterised by the relative humidity. Aqueous precipitation is best characterised by the time of wetness (TOW). TOW is the length of time that water is in direct contact with the coating surface as a separate liquid phase. Frequently, only data on the rainfall can be found because that is easier to measure than the TOW. However, TOW is not identical with the rainfall, because it includes all aqueous precipitation. The rainfall can serve only as a rough estimation of TOW. Water can be dissolved to greater or lesser extents in the polymer, and so can diffuse inside the polymer. Possible influences of water on coating ageing during normal use are listed in Table 3.6. **As can be seen, water has an accelerating effect on coating ageing in general.**

Table 3.6: Effects of water on coating ageing

The influence of water on the coating ageing		
	Effect	Response
1	supporting the decomposition of the O_2-molecule by UV radiation	acceleration of the photooxidation due to formation of the high reactive oxygen-containing radicals: OH· and OH_2·
2	decrease of glass transition temperature	accelerating of the O_2-diffussion inside the coating resin
3	hydrolysis of the polymer	molecule degradation and increase of the number of end double bonds acting as chromophores
4	cyclic mechanical stress by welling and shrinking	crack formation or delamination (occurring at sufficiently pre-weathered coatings only)
5	leaching of additives	loss of stabilization
6	supporting growth of algae and mildew	formation of pinholes

The first-mentioned factor is the most important. It was first identified in the weathering of TiO_2-pigmented polymers, which experience more intensive chalking in the presence of water. No chalking is observed during weathering in a dry atmosphere. What is the reason for this effect? The chemical decomposition of the rather inert O_2 molecule by UV photons to form special free-radicals together with water can be seen as the main cause of the effect promoted by water [35]. In the form of these free-radicals, oxygen is much more highly reactive. Metal ions that can easily switch between different valencies, such as those of titanium, iron, and copper, are necessary catalysts. Figure 3.38 shows the reaction between an oxygen molecule (O_2), a UV photon (hν), and a water molecule (H_2O) leading to hydroxide (OH·) and perhydroxy (HO_2·) free-radicals.

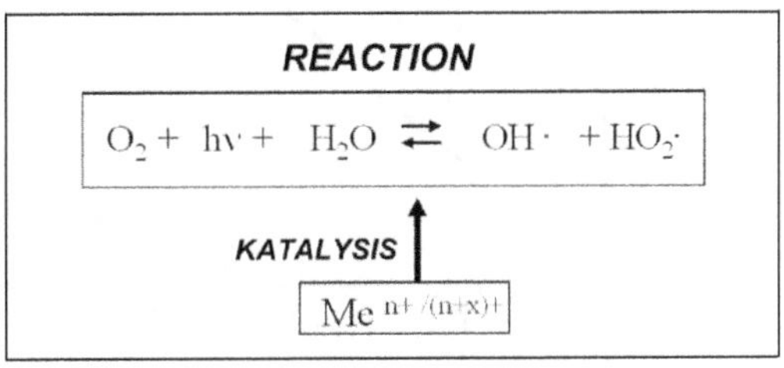

Figure 3.38: Highly reactive hydroxide and per-hydroxy free-radicals formed by conversion of an oxygen and a water molecule and initiated by an absorbed photon of global UV (hν)

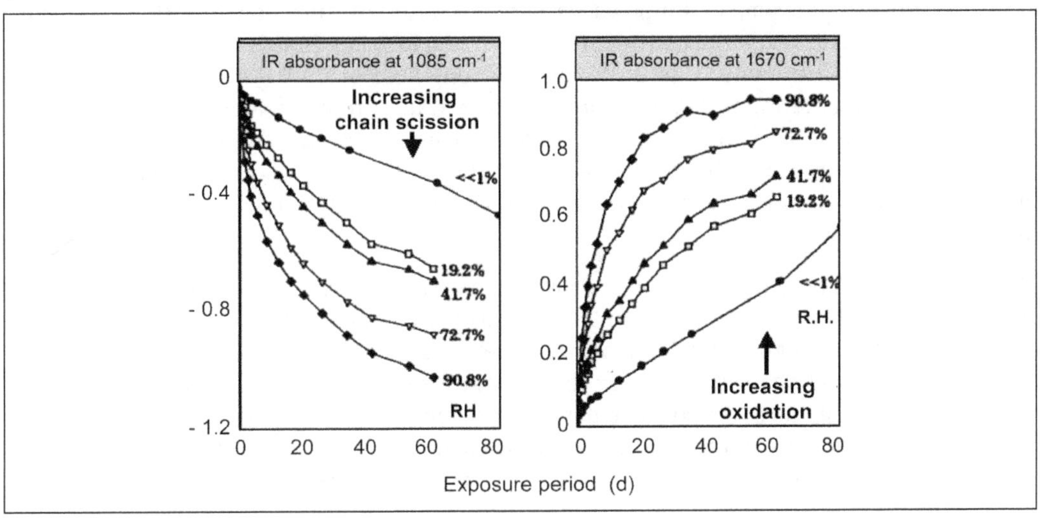

Figure 3.39: The influence of relative humidity (R.H.) on the increase in chain scission (decreasing IR absorbance at 1085 cm^{-1}) und the increase in formation of oxidation products (increasing IR absorbance at 1670 cm^{-1}) during artificial weathering of a melamine acrylic coating (adapted from [36]).

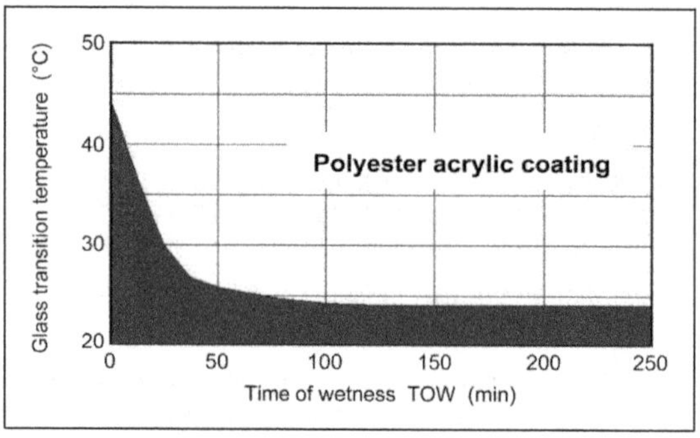

Figure 3.40: The glass transition temperature of a polyester acrylic coating as a function of the time of wetness (adapted from [37])

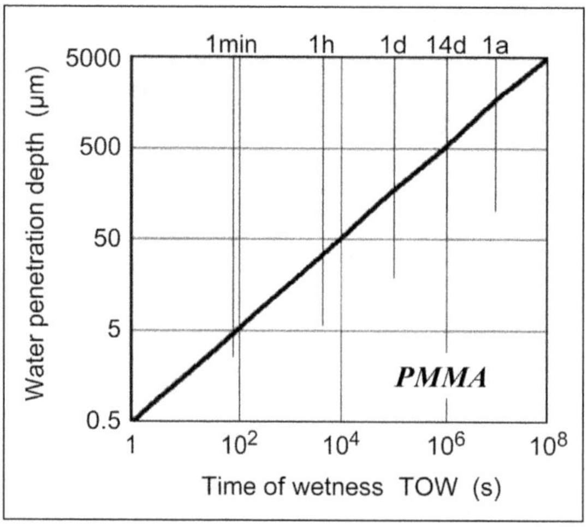

Figure 3.41: Penetration depth of water in cyclically wetted PMMA (adapted from [38])

Due to their high reactivity, the two free-radicals, formed by atmospheric oxygen and water, lead to more intensive photooxidation and therefore to faster coating ageing.

Figure 3.39 shows how moisture considerably promotes the photochemical ageing of a melamine acrylic coating during weathering. The higher the relative humidity of the test atmosphere, the faster photooxidation proceeds, as can be seen from the increasing number of chain scissions (measured as IR absorbance at 1085 cm^{-1}) as well as from the increasing content of oxidation products (measured as IR absorbance at 1670 cm^{-1}). In this case, increasing the relative humidity from <<1 % to 90.8 % produces a roughly fourfold acceleration of coating ageing.

Furthermore, it is known that water acts as a plasticiser in polymers to lower the glass transition temperature. The higher the water content, the lower the glass transition temperature, and the easier it is for oxygen diffusion to take place. The lower the glass transition temperature, the faster the oxygen molecules meet polymer free-radicals formed by UV photons. Consequently, water can accelerate coating ageing in this way, too.

Figure 3.40 uses the example of a polyester acrylic coating to show how quickly and steeply the glass transition temperature can fall after immersion in water. In this case, the glass transition temperature fell from 45 °C to 25 °C after just one hour of TOW. As frequently observed during outdoor weathering in Miami, Florida, sizable water drops remained on the surface of the test panels even after several hours of sunshine. This observation shows that UV radiation can act on surface areas which are completely saturated with water. The fastest coating ageing can be expected on surfaces of this kind because UV, heat, and water act highly effectively in concert.

Clearcoats can be expected to undergo the most intensive ageing if the hydroxide and perhydroxy radicals are formed inside the entire film. A prerequisite here is the ability of the clearcoat to dissolve a sufficient amount of water. For instance, water uptake of 1 % to 2 % can be reckoned with in the case of acrylic coating resins. However, the water needs sufficient time to penetrate far enough into the film. Investigations carried out at normal room temperature showed that it takes about 20 minutes of immersing to saturate a 30 µm-thick acrylic film with water, as seen in Figure 3.41. In a dry atmosphere, the water would need about the same time to leave the saturated polymer.

For clearcoats that absorb little or no water, the accelerating effect can only occur directly on the coating surface. Certain polymeric materials are known to have a tendency to hydrolyse. In this case, hydrolysis mostly supports the photochemical ageing.

Cyclical mechanical stress caused by cyclical wetting and drying are also listed in Table 3.6 as a further effect of water. Changes between high and low water content, for instance, due to changes between wet and dry weather scenarios, can lead to cyclical mechanical stress as a result of intermittent swelling and shrinking of the polymer. A prerequisite for cyclical mechanical stress capable of causing coating damage is the ability of the coating to take up enough water. Investigations described in [38] on PMMA roof panels capable of absorbing up to 3 % water have shown that alternating wetting and drying did not lead to any visible damage, even after a very high number of cycles. However, this result was only valid for non-weathered PMMA panels which had sufficient mechanical strength to resist such a stressful situation without damage. Repetitive cyclical wetting and drying stress acting on PMMA panels that had long been exposed to natural or artificial weathering led to the formation of cracks. In other words, only after the polymer had lost considerable mechanical strength due to weathering, was it possible for cyclical mechanical stress in the form of alternating wetting and drying to cause damage.

Non-weathered coatings with enough initial mechanical strength do not seem to suffer degradation from cyclical changes in water content. Similarly, cyclical stress by water, e.g. in the form of regular changes in wetting and drying, could also serve as a tool for shortening accelerated testing - not because it would accelerate coating ageing but because it would render the state of ageing, e.g. formation of cracks, visible earlier.

It has been reported that certain polymer materials have a retarding effect on water. Polyoxymethylene is a typical example; it exhibits the longest service life for weathering involving permanent wetting of the irradiated sample surface. There are no plausible explanations for this phenomenon. However, it cannot be ruled out that, with regard to a specific coating property, the seemingly positive effect of water is due to other ageing effects which do not show up as negative. The ageing behaviour of window frames made from TiO_2-pigmented PVC, a common building material nowadays, is typical of such a case. As is well known, this product tends to yellow in dry atmospheres, but retains its white colour quite well in humid ones. This is attributed to competition between two photochemical reactions which occur one after the other during weathering. The first reaction is the photodissociation of PVC, with cleavage of HCl to leave behind conjugated double bonds, which are responsible for the yellow colour. If sequences of more than 8 conjugated double bonds are formed, the PVC will undergo strong yellowing. The rate of this partial reaction is water-independent. The subsequent second reaction is photooxidation of the conjugated double bonds, which bleaches the PVC. Double bonds, which act as chromophores, are preferably attacked by oxygen. As mentioned earlier, the reactivity of oxygen is promoted by water. In other words, the rate of ageing is water-dependent. Whether the window frames exhibit yellowing or bleaching depends on the reaction rates. If the photodissociaton proceeds faster than the photooxidation, the frames will turn yellow. If photooxidation is the fastest process, the conjugated double bonds will not reach the number of sequences necessary for rendering the yellowing of the PVC visible.

According to the equation in Figure 3.38, sufficient hydroxy and perhydroxy free-radicals are formed in the presence of water, TiO_2, UV radiation and oxygen, to make photooxidation of the conjugated double bonds the fastest step. Note that both reactions lead to degradation of the PVC and to chalking promoted by water! However, chalking of window frames is not as noticeable as yellowing. The example of white PVC window frames readily shows that water can extend service life due to a hidden water-promoted ageing effect. Nor can it be ruled out that water has a protective action on certain coatings as well, especially those containing TiO_2.

3.2.4 Ambient climate

So far, we have dealt with the most important factors of weathering stress. Global radiation, atmospheric oxygen, heat, and water, and their influences on coating ageing, have mainly been characterised in isolation from each other. As already discussed before, the complex action of the aforementioned weathering stresses must not be neglected. Synergies can boost the action of certain weathering stresses and diminish others. The corollary of this is that weathering stresses influence each other. It is often observed that a particular weathering stress factor only comes into play in the presence of the other. For instance, if the weathering stress factor heat is absent, as in alpine regions or at high flight altitudes, no substantial ageing will occur in spite of high UV irradiance. During cool weather episodes, too, coating ageing may be fully neutralised.

Exposing coatings in strongly arid desert regions often leads to barely discernible ageing, because the necessary moisture is absent in spite of high yearly radiant exposure. It can be inferred from these examples that the weathering stress factors always act in complex ways and that all the weather situations which occur at the location of normal use must be considered when real weathering stress is being simulated. The examples also show that weathering stress can vary enormously from locality to locality. For the purpose of designing a method for accelerated testing that reflects real weathering conditions, the terms ambient climate and outdoor climate are used in the literature. The terms mean the same. The rest of this paragraph is devoted to the natural climate as the natural form of the ambient climate. The natural climate is predominantly determined by the incident solar radiation, the degree of latitude, the altitude, the distribution of land and sea, ocean currents, building culture, and vegetation. In line with this definition, the ambient climate is usually described by the climatological data listed in Table 3.7. The most important weathering stresses, such as radiation, heat, and water are more or less adequately described by the measurable parameters listed. It is usual to take a period of one year as the basis. When designing the most convenient method of accelerated testing, worst case weather scenarios, as discussed in Chapter 2.4, can be derived from these ambient climate data. Furthermore, these data measured during outdoor weathering can serve to characterise and assess the received weathering stress.

Table 3.7: Basic climatological data for characterising the ambient climate during outdoor weathering

Climatological data	UOM*
Global radiation (GI) (300 to 2500 nm) global UV (GIUV) (300 to 400 nm)	MJ/m^2
Ambient temperature	°C
BST or BPT	°C
Relative humidity	%
Amount of rain	mm
Time of wetness (TOW)	h

**Unit of measurement*

Since there are huge geographical differences in terms of the ambient climate, there have been many attempts to catalogue the world's climate with a view to arriving at a standardised basis for describing weathering stress in regard to the intended normal use for outdoor weathering. The ISO 877 classification of

Table 3.8: Classification of ambient climates according to ISO 877

	Ambient climate	Geographical reference
1	tropical rain climate	tropical rain forest tropical savannah
2	dry climates	steppe-tropical/subtropical steppe in middle degrees of latitude desert-tropical/subtropical desert in middle degrees of latitude
3	humid, medium warm climate	Mediterranean, subtopic/dry summers humid subtropical/warm summers ocean climate/could summers
4	humid microthermic climate	humid continental/warm summers humid continental/could summers sub-arctic
5	high land climate	
6	polar climate	

ambient climate shown in Table 3.8 is but one typical example. When natural stress is used to design a time-compressed type of accelerated testing (discussed in Chapter 2.5), the most suitable climatic regions are those in which weathering stress factors, such as global radiation, heat, and water, occur to natural extremes, and the corresponding worst case weather scenarios occur the most frequently. For most coatings, outdoor weathering can be expected to be most efficient if performed in tropical rain climates.

3.2.5 Coating surface micro climate

Having discussed the weathering stress factors heat and water in Chapter 3.2.2 and Chapter 3.2.3 in isolation, some consideration of their mutual action on the coating surface is needed. The ambient climate, as discussed in Chapter 3.2.4, describes only the environment in which the coatings are exposed. Therefore, the ambient climate can be seen as independent of the type and properties of the coating to be tested. However, photooxidation that leads to coating ageing predominantly occurs at the interface of the coating's outermost micrometers and the sheet of gaseous or liquid molecules from the ambient atmosphere that is in direct contact with the coating surface. On this surface, global radiation, oxygen, and water, react with the coating molecules. It is easy to realize that the rate of photooxidation, which predominates here, depends on the conditions at this interfacial area. Since this area involves only a few micrometers of the ambient atmosphere, its climate is generally termed an interfacial or micro climate. In terms of coatings, the interfacial area of interest can be considered to be almost identical with the surface of the coating. For this reason, the term coating surface micro climate shall be used below.

Let us start by looking at the surface temperature. This depends first on the irradiance of the incident radiation. Outdoors, the source is global radiation while, for artificial weathering, it is a lamp. The higher the irradiance, the higher is the surface temperature. Second, the surface temperature is determined by the absorbance und reflectance of the coating, which are typical coating properties. The higher the absorbed radiation, the higher is the surface temperature. In other words, dark-coloured samples reach higher temperatures than light-coloured ones. Third, the surface temperature is influenced by energy transport processes on or through the irradiated test panel which is trying to dissipate the absorbed radiation energy (for instance, wind velocity, or cooling of the rear side). This means that the coating surface micro climate does not only depend on the ambient climate, but also depends heavily on the coating's properties and special exposure conditions. In the event that no direct radiation acts on the coating surface, for instance, under clouds, the ambient climate and coating surface micro climate can be almost identical. Therefore, these two types of climate can correspond to each other as well as deviate extremely from each other (for instance during unhindered irradiation by the sun). The appearance of high weather-related differences in the surface temperatures can be best explained by the thermal equilibrium between received and donated radiation energy. Figure 3.42 explains the influences on the surface temperature arising from this equilibrium.

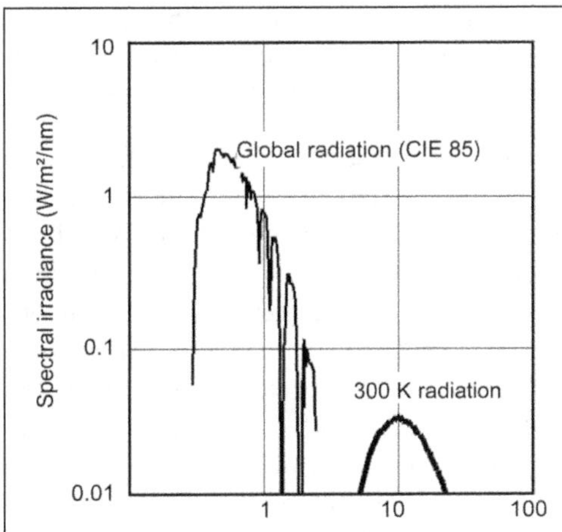

Figure 3.42: Spectral irradiance of global radiation (according to CIE85) and the radiation emitted by a 300 K black body (adapted from [39])

This Figure shows a section of the electromagnetic spectrum comprising global radiation (according to CIE 85, Table 4) and the radiation emitted by an ideal black body at 27 °C.

$$\lambda_{E\text{-}MAX} = 2900/\,T$$

Figure 3.43: Wien's law ($\lambda_{E\text{-}MAX}$: Wavelength at the irradiance maximum (μm), T: Absolute temperature)

$$E = \sigma \cdot T^4$$

Figure 3.44: The Stefan-Boltzmann law (E: Emitted energy, σ: Stefan-Boltzmann constant, T: Absolute temperature)

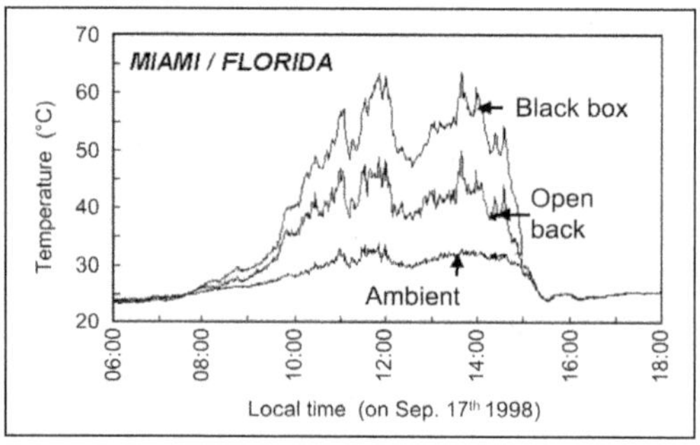

Figure 3.45: Comparison of the ambient temperature with the surface temperatures of panels with different back-side conditions during outdoor exposure in Miami, Florida, on 17.09.1998 (adapted from [39])

According to Wien's law, presented in Figure 3.43, the position of the irradiance maximum is determined by the surface temperature of the respective radiation source. Deriving from the sun's temperature of some 6500 °C, the irradiance maximum of the global radiation can be found at a wavelength of around 420 nm, where it reaches about 1100 W/m². Assuming that about 60 % of the energy of the incident solar radiation is lost by reflection, convection, or conduction, the surface of the irradiated test panel would warm up to 27 °C. At this temperature, the test panel emits the residual 460 W/m² in the form of radiation. As can be seen in Figure 3.42, the radiation emitted by an ideal black body has an irradiance maximum at 10 μ m, which is inside the thermal infrared range. If we presume, as an example, that the coating is an ideal black body and absorbs 100 % of the incident global radiation without any loss of energy by convection and conduction, the test panel surface would attain a temperature of about 100 °C in line with the Stefan-Boltzmann law.

In this case, all of the received 1100 W/m² will be emitted in the form of radiation having an irradiance maximum at about 8 μm. As reported, the surface of black-coloured coatings frequently reaches 80 °C outdoors near the equator. Therefore, let us look at the coating's surface temperature first. As measured during conventional outdoor weathering in Miami, Florida, at noon on a clear day in June, the temperature of test panels of black-coloured automotive coatings reached 60 °C. However, only 45 °C was measured on test panels of white-coloured counterparts simultaneously exposed among the black ones. The reason for this temperature difference lies in the colour-related differences in absorbance and reflectance. Differently-coloured coatings ranked in between, as a function of their luminosity.

As furthermore observed, common changes in the wind velocity cause the surface temperature to vary by a few °C, due to changes in convection. In contrast, complete confinement of the air circulation at the coating surface, e.g., by exposing the test panels mounted on a black box, causes the surface temperature to rise by about 15 °C almost irrespective of colour. Another example is shown in Figure 3.45. Here we see the surface temperatures measured on identical test panels simultaneously exposed to conventional outdoor weathering (open at the back) as well as on the black box. The measurements were carried out on a September's day, in Miami, Florida. The measured surface temperatures were plotted 6 a.m. to 6 p.m., together with the respective ambient temperature. The highest differences between the ambient climate (represented by the ambient temperature) and the coating surface micro climates occurred at the time of the sun's highest angle of elevation between 10 a.m. and 3 p.m., i.e. at the highest irradiance of the global radiation.

The influence of changing irradiation stress is also shown by a further example. It describes observations that were also made during conventional outdoor weathering of black-coloured automotive coatings under a partially covered sky in Miami, Florida. The observation was made at noon on a day in June. It was found that the change from the state of highest irradiance (the sun's rays reached the specimen unhindered by clouds) to the state of lowest irradiance (the sun's rays were prevented from reaching the specimen by a cloud) lowered the surface temperature from 60 °C to 35 °C after only a few seconds. In other words, the surface temperature decreases to virtually ambient temperature if direct irradiation by the sun is interrupted. These discussions clearly show that ambient temperature can serve only as a descriptor of the ambient climate. For the most part, the specimens' surface temperature does not match the ambient temperature. Since weathering is mainly a surface effect, the coating's surface temperature has more influence on coating ageing than has the ambient temperature.

Since precise data are needed to adequately describe the thermal stress during accelerated testing, monitoring of the coating surface temperature would be desirable. However, it is not feasible to measure the surface temperature of each specimen during weathering. Instead, the so-called black panel temperature (BPT) or black standard temperature (BST) is usually monitored. Standardised instruments for measuring these are available. They are usually black panel thermometers or black standard thermometers. These two types of thermometers differ in their design details. The monitored data can differ by a few °C, which is mostly within the commonly accepted tolerance of ±3 °C.

For better reproducibility and conformity of weathering tests, BST should be preferred worldwide. The sensors in both types of thermometer are special black panels of high radiation absorbance (93 to 95 %). The measured BPT/BST is considered to be the highest temperature that specimens may reach on exposure to weathering. Therefore, BPT/BST allow only approximate statements about the actual surface temperature during weathering. Due to the aforementioned high dependence of the surface temperature on the respective coating surface properties, the nature of the substrate, and the type of mounting, BPT/BST measured in this way can never be identical with the actual values, even if mounted among the test panels during weathering. Nevertheless, BST/BPS can contribute extensively to an assessment of the thermal stress during accelerated testing and to comparisons of different or repeated weathering tests. During outdoor weathering, BST/BPS are commonly exposed at different exposure angles facing the equator, these being mostly identical with the angles used for exposing the test panels. BPT/BST also deliver important information about the ambient climate in addition to the daily average irradiance data usually published. The days can differ in the types of cloud, for instance, whether they are discrete or homogeneously distributed as mist. Such a difference can yield differing values for daily UV radiant exposure, in spite of identical global radiation. Such additional information can be derived from the data of the maximum values BST_{max}/BPT_{max}. Perhaps for the same reason, global radiation and BPT_{max} do not follow the same course during the year, as shown in Figure 3.49. The strongest deviations can occur in the summer months, where Miami, Florida, is known to have the highest cloud coverage.

In artificial weathering, the function of BST or BPT is often to act as a control for the weathering devices. In basecoat/clearcoat paint systems, an additional interfacial area between the clearcoat and the basecoat must be considered, at least in terms of temperature. As clearcoats are mostly highly transparent in the VIS and NIR ranges, the incident global radiation passes almost unhindered through the clearcoat, and is predominantly absorbed by the topmost micrometers of the basecoat. In contrast to the global radiation, the thermal infrared radiation emitted by the warmed basecoat is almost completely absorbed by the clearcoat on its way back to the clearcoat surface. As result, the interfacial area between clearcoat and basecoat can reach a higher temperature than that of the clearcoat surface.

Now, let us turn to moisture as the second most important part of the coating surface micro climate. As mentioned before, the relative humidity proves to be the determining climatic parameter for characterising the action of moisture dissolved in the air. The absolute water content, as expressed, for instance, by the dew point, should be identical for both the ambient

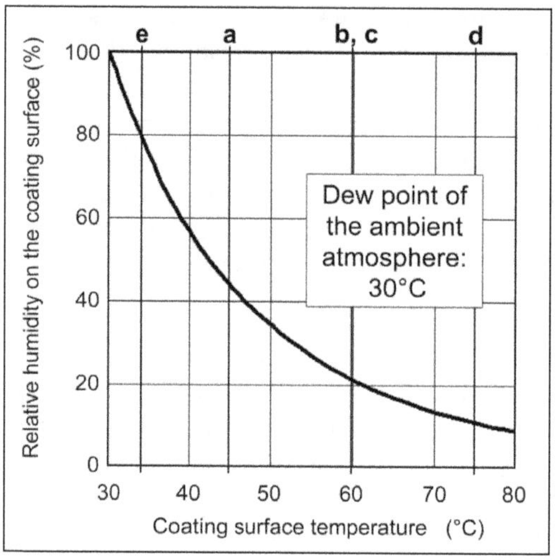

Figure 3.46: Relationship between coating surface micro climate and ambient climate, shown here for an ambient atmosphere with constant dew point (30 °C) for five typical weather scenarios that frequently occur at noon on summer days in Miami, Florida (test panels horizontally exposed in each case)
a: white coating/open back/clear sky. b: black coating/open back/clear sky, . c: white coating/black box/clear sky. d: black coating/black box/clear sky. e: black or white coating/open back/a cloud in front of the sun

atmosphere and the atmosphere immediately on the coating surface. However, the relative humidity immediately on the coating surface strongly decreases with increasing surface temperature. So, the relative humidity on the coating surface depends not only on the water content of the ambient atmosphere, but also on the coating's surface temperature, and therefore, also on the irradiance. This means that, in general, the higher the irradiance, the lower is the relative humidity on the irradiated coating surface. The following example demonstrates the surface climate as a function of the weather situation for Miami, Florida, at noon on a typical summer day characterised by an almost constant ambient climate. Figure 3.46 shows how the coating surface micro climate changes between weather episodes with and without direct irradiation for five typical exposure scenarios (a) (b) (c) (d), and (e), with allowance for the test panels' colour and backside conditions.

The relative humidity on the coating surface influences how fast and how extensively the coating loses the water that remained on the surface after a rain shower or was taken up by the polymer matrix. As discussed for the weathering stress factor water in Chapter 3.2.3, moisture mostly accelerates coating ageing. For instance, taking the example of Figure 3.46, the more humid coating surface micro climate of scenario e (34 °C/80 % R.H.), with 50 % cloud cover, can be expected to lead to faster coating ageing than the drier scenario b (60 °C/20 % RH.), for moisture-sensitive coatings in spite of higher irradiance and higher surface temperature. As a function of coating type, one or other of the aforementioned exposure scenarios could result in faster coating ageing and, therefore, represent the worst case weather scenario.

As already discussed in Chapter 3.2.3, water in gaseous form, dissolved in the ambient air and expressed by the term relative humidity, is only one form of this weathering stress factor. Water in liquid form, arising from aqueous atmospheric precipitation, such as rain and fog, or from condensation on the surface, is generally seen as the more exacting form. A coating can absorb water up to its saturation point. As the kinetics discussion in Chapter 3.2 revealed, the diffusion rate of water as it passes from one phase to the other depends on the difference between the water contents of the phases concerned. On its way into the polymer matrix, water will reach saturation point faster if the water-providing phase in touch with the coating surface has higher water content. The content of water dissolved in ambient air only reaches about 5 % at 100 % R.H. and 35 °C. Aqueous precipitations consist of about 100 % water.

Extensive studies of the influence of water on PMMA plate material [18] have shown that the higher the water content inside the polymer, the faster is the photochemical ageing. These studies demonstrated that the highest ageing rate would be attained if the specimen's surface were continuously wetted during weathering. Aqueous precipitations contribute most to a coating's water uptake. Nocturnal rain showers or wetting by condensation determine the water content with which the coating starts to face the sun in the early morning hours. The number of hours wherein water is present in liquid form on a surface is denoted by the term TOW (time of wetness), which is frequently used as a descriptor of both the coating surface micro climate and the ambient climate. Figure 3.47 shows how TOW is influenced by the ambient climate. The humid subtropical climate of Miami, Florida, and the desert climate of Phoenix, Arizona, clearly differ in the yearly change in TOW. As can be seen, TOW reaches 60 % in Florida, but only 16 % in Arizona. TOW can therefore also serve to differentiate ambient climates.

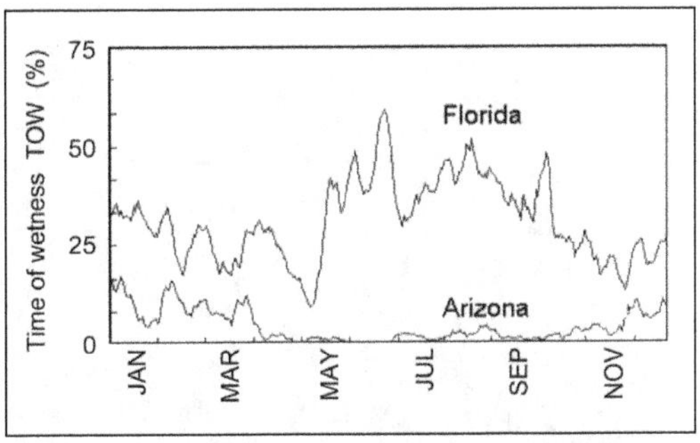

Figure 3.47: TOW in the course of the year, measured daily in Florida and Arizona (adapted from [9])

Dew contributes most to TOW. Comparison of annual average values shows that test panels are wetted for about 380 hours by rain but by 4370 hours (more than ten times as much) of dew outdoors in Miami, Florida. This ratio is similar to that in Phoenix, Arizona, where the test panels are wetted for 43 hours by rain and 219 hours (about five times as much) by dew. Given the particular meaning of dew, a further discussion of this type of stress by water seems appropriate.

As discussed in the context of Figure 3.42, the test panel surface emits energy in form of thermal infrared according to its temperature, also after irradiation has ceased. Thermal equilibrium is established between the thermal energy emitted to and received from the environment. At night, especially under clear skies, the thermal energy emitted by the clouds is absent. If, additionally, there is an interruption in the transport of energy by radiation, convection, or conduction from other sources of warmth to the test panels, the coating surface temperature can quickly fall by 10 °C relative to the ambient temperature. As a result, the dew point temperature can be reached, and dew starts to form. The earlier the formation of dew begins, the longer is the TOW and the possibility that the coating will reach saturation with water. As already discussed in more detail in Chapter 3.2.3, the destructive action of global radiation can increase enormously on the coating surface in the presence of water due to synergies. The water-related acceleration of coating ageing is generally a result of the mutual action of UV radiation and water.

As described in more detail in Chapter 3.2.3, the hydroxide (OH•) and perhydroxy (HO2•) free-radicals, formed by reaction of oxygen, UV photons (hν), and water were found to be the main reason for accelerated coating ageing. Therefore, the duration of simultaneous action of UV radiation and water on the coating surface (TOW+U) must be given more attention when proper weathering tests methods and test devices are being designed. The aforementioned free-radicals can be formed only in the presence of UV radiation. Without UV radiation, the stress factor water cannot unfold its destructive action. This means that the number of hours during which water plus UV radiation are simultaneously present on the coating surface (TOW+U) is more influential on coating ageing than the commonly monitored TOW. As is known, TOW reflects the hours in which the surface was wet, but ignores the presence or absence of UV radiation. TOW+U proves to be a further very important feature of the coating surface micro climate. The photograph shown in Figure 3.48, which was taken during conventional outdoor weathering in Miami, Florida, helps to clarify the importance of TOW+U. It shows test panels of 16 different black automotive coatings, mounted almost horizontally on a common weathering rack. The photograph was taken at noon on a summer's day, about 5 hours after an intensive rain shower. The sky was covered by clouds to about 50 %. The photograph shows first that, after five rain-free hours during which there was more or less direct irradiation, an astonishing amount of water was still present on the surfaces of the test panels. Second, it was always the case that only a part of the test panel surfaces was covered with water in the form of more or less extensive puddles. Third, the specimens differed in the size of the water-covered areas on the test panel remaining from the last rain shower. As can be seen, while the surface of one test panel was still some 50 % covered, the surface of another one was already completely dry.

Figure 3.48: Photograph of test panels with different black-coloured automotive coatings taken after a rain shower followed by several hours of sunshine during outdoor exposure in Miami, Florida, in summer 2002

Furthermore, it was observed later on the same day that a few of the test panels remained wet throughout the day due to further, mostly brief rain showers, while other ones dried out completely in the interim. Since the water puddles observed on the test panels did not prevent UV from reaching the coating surface to any great extent, radiation simultaneously acted on water-covered and uncovered surface areas. In other words, UV simultaneously acted on differently water-stressed surface parts of the same test panel. Like TOW, TOW+U depends on the surface properties of the coatings being exposed. In this example, the relationship to the coating properties is revealed by the different ratios between wet and dry surface areas, as well as the duration of the presence of liquid water on the test panels. In spite of having the same colour, all test panels shown in Figure 3.48 exhibited considerable differences in TOW+U. Therefore, it can generally be stated: “one black is not the same as another black”. Unfortunately, no data on the TOW+U for outdoor weathering are available. However, TOW+U can be considered when proper artificial weathering tests are being designed. Data for TOW are sometimes made available by the operators of outdoor weathering sites. However, they can only be seen as a rough approximation of the actual conditions. They are commonly measured with standardised instruments according to ASTM 684 [40] and so are valid only for the standardised measuring conditions described therein.

In order to be able to characterise a weathering test as comprehensively as possible, staff performing weathering tests should at least obtain the climatological data represented in Table 3.7. Data for hours of sunshine, which are sometimes found in older literature, should no longer be used. The strong variations in global radiation, in the course of the day, month, or year, renders this climatological parameter useless.

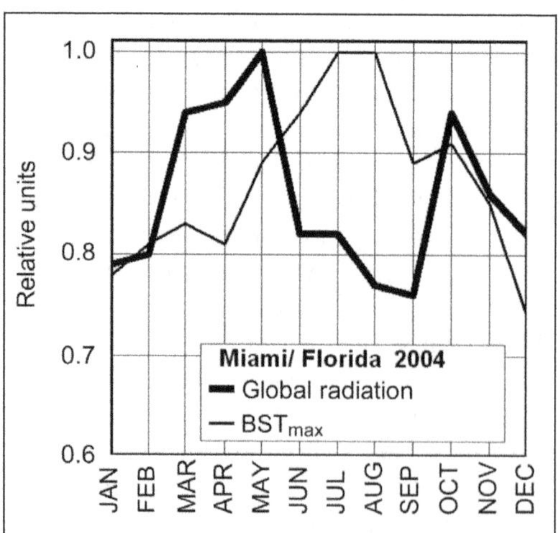

Figure 3.49: Monthly radiant exposure to global radiation and monthly average values of the maximum black standard temperature (BST_{max}) measured at 26° south, in Miami, Florida, in 2004 (adapted from [30])

3.2.6 Weathering stress factor acid atmospheric precipitation

Stress by weathering that merely contains stress by irradiation, heat and moisture does not necessarily represent the worst case weather scenarios for certain types of coatings.

Special anthropogenic influences, arising especially from burning fossil fuels to produce electrical energy, traffic, industry, and domestic heating, generate additional environmental stress by aggressive air pollutants and acid atmospheric precipitation. Acid rain, acid dew, and acid fog are known to cause considerable coating damage, even in relatively harmless climatic conditions. The following examples are intended to illustrate that acid atmospheric precipitation belongs to those weathering stress factors which predominate in industrial conurbations.

Figure 3.50 shows the loss of gloss measured on five different clearcoats after two years of outdoor weathering performed simultaneously at five different locations. As can be seen, sunshine is involved in all worst-case weather scenarios. Although, in this case, the tropical rain climate of North Australia was the expected worst case weather scenario of all of the tested clearcoats, only clearcoat #5 fulfilled this expectation. Surprisingly, most of the clearcoats (#1, #3, and #4) revealed the environment of Denmark to provide their worst case weather scenario. The presence of acid atmospheric precipitation due to greater industrialisation is clearly the reason for the faster coating ageing in Denmark in spite of its relatively benign climate.

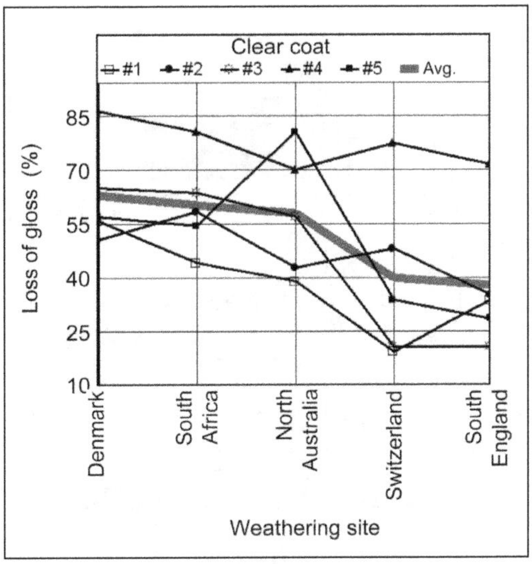

Figure 3.50: Loss of gloss by five different clearcoats (#1 to #5) weathered outdoors at different weathering sites (exposure period: 2 years) (adapted from [42])

In another example, the results of outdoor weathering carried out in Miami, Florida, and Sanary, Southern France, were compared. The tests served to examine the influence of certain light stabilisers on the ageing behaviour of PE film [43]. Meteorological data were unable to explain the differences in the weathering results that were obtained. Another good example is the results of outdoor weathering performed on identical series of 12 different types of pigmented coatings at three weathering sites located in different climatic regions [44]. After one year of exposure, the loss of gloss and degree of chalking were determined.

Table 3.9 shows that the most important ageing criteria for pigmented coatings differ as a function of the chosen weathering site. The three weathering sites were ranked in terms of the degree of damage that was measured after exposure. The first thing to notice is that the rankings varied with the ageing criterion applied. For loss of gloss, Aquaba, Jordan, proved to be the most damaging region. As regards chalking, Lake Constance proved to be more damaging than Aquaba in spite of its higher irradiance and temperature. The proximity of Lake Constance to the industrial conurbation of Basle (Switzerland) and Rheinfelden (Germany) is obviously the reason for this result. Evidence of the occurrence of acid atmospheric precipitation in central Europe was obtained by several governmental institutions.

The results of rain measurements taken above Mt. Brocken in Germany can be regarded as representative. Whenever rain clouds came into view over the 1000 m-high mountain, balloons were sent up to extract rain water directly from them. Figure 3.51 (page 60) shows the frequency distribution for the acidity (pH values) of the collected rain water samples in 1992, 1996, and 2000.

Table 3.9: Weatherability rankings for outdoor weathering as a function of the type of ambient climate and the chosen ageing criterion (adapted from [44])

	Rank 1 (strongest damage)	**Rank 2 (medium damage)**	**Rank 3 (least damage)**
Loss of gloss	dry-hot climate (Aquaba)	humid-hot climate (Florida)	moderate climate (Lake Constance)
Chalking	humid-hot climate (Florida)	moderate climate (Lake Constance)	dry-hot climate (Aquaba)

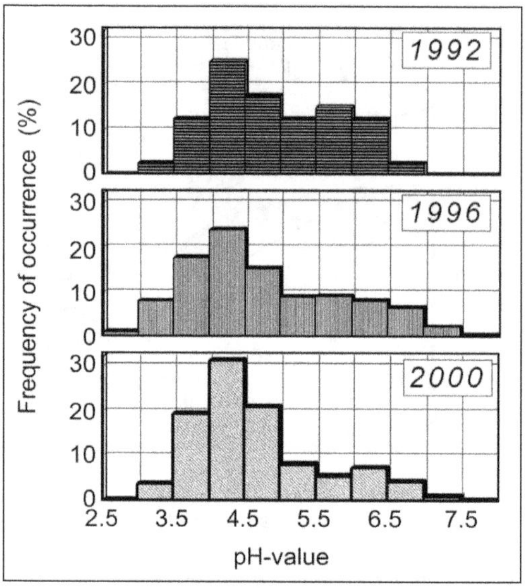

Figure 3.51: Frequency distribution of the pH value in cloud water at Mt. Brocken, Germany, in years 1992, 1996 and 2000 (from [45])

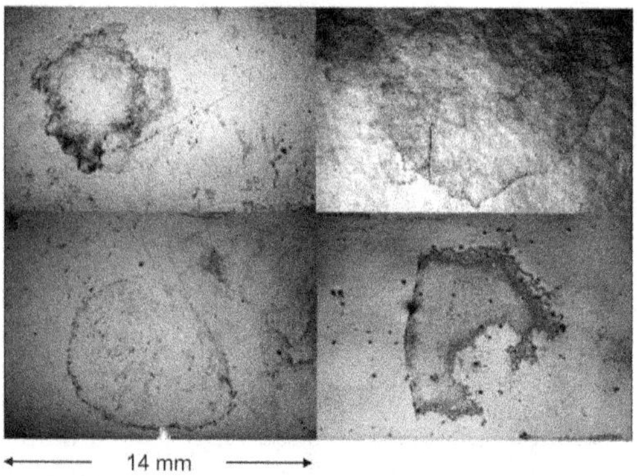

Figure 3.52: Typical acid etching on the surface of automotive coatings after outdoor weathering in Jacksonville, Florida

Aqueous precipitation with pH values <5.6 is acidic. Therefore, according to this Figure, 70 % to 80 % of the rainfall can be ranked as acid atmospheric precipitation. A pH of 2.5 was the lowest reading and represents highest content of acid. The next example comes from the USA. The typical acid etching observed on automotive coatings during outdoor storage of imported cars in the harbour area of Jacksonville, Florida, since the early 1980s is clear evidence of the destructive effect of acid atmospheric precipitation. Acid etching takes the form of pit marks up to 1 cm wide and 1.5 μm deep. These pit marks appear as spots on the coating surface. The spots cannot be removed by washing or polishing. Figure 3.52 shows photographs of typical acid etching that were taken after outdoor weathering of automotive coatings in Jacksonville, Florida.

Extensive studies [45] confirmed that this typical damage was caused by acid atmospheric precipitation in the form of acid rain, acid dew, or acid fog, in combination with sufficiently high UV irradiance, heat and moisture. Evidence of the presence of acid atmospheric precipitation was provided by a study carried out in the harbour area of Jacksonville, Florida, in 1988 [46]. Every rainfall that occurred in this period was registered and analysed. The results are presented in Table 3.10. There were considerable differences in the quantity and composition of rain collected on rainy days.

All the rain samples were classified as acidic. The sulphate content was also remarkable. Some of the other ingredients may have stemmed from sea salt due to the proximity of the ocean. The pronounced variations in pH values between 3.61 and 5.51 and the daily precipitated acid between 0.003 and 4.5 μg/m² H^+ ions are striking. Marked deviations in acidity are a typical feature of acid atmospheric precipitation. As has been found in other industrial regions of the globe, acid atmospheric precipitation does not occur continuously, but rather is sporadic and random. In other words, its timing, acidity, and quantity obey the laws of chance and so are not predictable. Since dew and fog occur more frequently than rain, it can be reckoned that the actual stress by acid atmospheric precipitation in Jacksonville, Florida, is higher than presented in Table 3.10. Dew and fog have a higher acid content than rain. Dew is frequently ten times as acidic as rain.

Coating damage is often also the result of the action of certain air pollutants. As described in Chapter 2.2, some of the corrosion tests were carried out in SO_2 containing test atmospheres. **The coating damage observed was obviously the result of an overly excessive SO_2-concentration, often 1000 times the value found in reality.** Common aggressive air pollutants are only found only in very low concentrations in the atmosphere (<1 ppm of ozone, nitroxides, sulphur dioxide, and chlorine). Extensive studies show that aggressive air pollutants have less of an

Table 3.10: Quantity and chemical composition of rain precipitated in the harbour area of Jacksonville, Florida, during the period from 1stJune to 2nd July 1988 (adapted from [46])

Date	Amount of rain (mm)	pH-value	Precipitated amount of the rain's ingredients (mg/m²/rainy day)							
			Ca^{++}	Mg^{++}	Na^{+}	K^{+}	Cl^{-}	NO_3^{-}	SO_4^{--}	H^{+}
June, 3rd	15.8	3.75	5.2	4.3	33.7	3.5	67.2	35.8	84.8	2.8
June, 4th	0.2									
June, 5th	0.5									
June, 6th	5.8	4.10	0.9	0.2	1.8	0.4	3.2	11.2	17.3	0.47
June, 8th	11.9	4.23	4.5	0.6	2.4	0.3	3.9	7.6	32.2	0.7
June, 9th	0.8	5.51	1.5	0.1	0.8	0.1	1.3	1.7	2.1	0.003
June, 10th	7.4	4.12	0.4	0.2	1.0	0.2	1.8	5.9	16.3	0.55
June, 12th	< 0.2									
June, 26th	18.3	3.61	8.5	3.5	7.6	1.4	18.5	57.7	200	4.5
June, 27th	23.9	4.00	2.2	0.5	1.5	0.5	3.6	25.6	61.9	2.4
June, 30th	< 0.2									
July, 1st	8.1	3.67	5.2	0.7	3.9	0.7	11.4	26.5	77.5	1.7

influence on coating ageing than the common variations of global radiation, temperature, and humidity, during outdoor weathering [47].

Acid atmospheric precipitation is the result of the aforementioned gaseous air pollutants. In the atmosphere, the lifetime of sulphur dioxide and nitroxides is limited to only a few days. Catalysed by global radiation, these gases react with oxygen and water to form the acids H_2SO_4, HNO_3 and HCl, which, diluted by atmospheric water, are deposited as acid rain, dew, or fog on the surface of coatings outdoors.

The damaging action of acid atmospheric precipitation is predominantly attributed to the special properties of sulphuric acid. In contrast to HCl and HNO_3, sulphuric acid is unable to evapourate under the climatic conditions of normal use. Any H_2SO_4, precipitated as dew in the early morning hours becomes concentrated by the action of heat in the course of the day up to a figure of around 70 %, the precise value depending on the surface temperature and the ambient relative humidity. The acid droplets concentrated in this way would remain on the coating surface unless they were removed by washing or showers of neutral rain. The acid therefore has plenty of time to attack the coating. Acid is known to boost the destructive action of global radiation due to synergy. Figure 3.53 reveals these synergistic effects for the example of automotive coatings exposed to artificial weathering.

In this example, two identical series of seven different types of automotive OEM basecoat/clearcoat systems were included in the test. During weathering, the test panels of both series were sprayed once a day with artificial acid dew (pH 2), while one of the two series was permanently sheltered against the action of UV radiation during weathering. In other words, both series were regularly attacked by acid, but only one series was exposed to a combination of acid and UV radiation. After 40 days of exposure (173 MJ/m² UV), the loss of gloss was measured. The diagram shows that

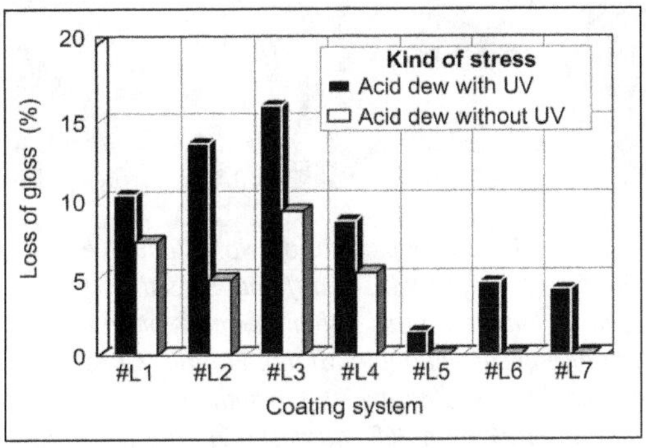

Figure 3.53: Synergism of global UV and acid dew for the example of automotive coatings (from [48])

all the coatings suffered from extensive acid-related degradation in the presence of UV radiation due to synergy. As already discussed in Chapter 3.1.4, acids can penetrate into the coating. The accelerated coating degradation due to UV radiation applied after an acid attack can be attributed to acid-induced loss of light stabilisers.

3.2.7 Biological stresses

Stress due to biological sources, such as the excrement of animals (birds, aphids, etc), resins and pollen from trees, fungi, and algae, also occurs in the natural environment, and so belongs to the weathering stress that occurs during normal use of coated products. Permanent parking of cars below trees nowadays counts as normal use. Studies of the action of special organic deposits on automotive coatings reveal that the effects of biological attacks were mostly synergistically boosted by global radiation. That has been proved with aphid excrement which was naturally deposited on test panels and subsequently exposed to artificial weathering.

During weathering, half of each test panel surface was sheltered against irradiation. After 28 days of exposure, the test panels were washed, polished, and visually assessed. Only those surface parts which were simultaneously attacked by aphid excrement and UV radiation revealed striking spots that could not be removed by washing and polishing. In contrast, the residues of the aphid excrement on the sheltered surface parts did not lead to coating damage and were easily removed by water. Resistance tests on outdoor coatings against biological attack should therefore only be carried out in the form of weathering that allows for possible synergies with other weathering stress factors.

The next example supports this assertion. Starting in the mid-1990s, a special type of damage came in for higher attention. Certain types of new automotive coatings exposed outdoors in Miami, Florida, (more precisely: Okeechobee Road weathering site, 26°N, 81°W) exhibited pronounced surface damage predominantly in the form of pinholes after just a few summer months. Pinholes are craters up to 1.5 µm deep on the coating surface. Extensive algal and mildew growth observed on the exposed specimens was blamed for this damage. To prove this, extensive studies were carried out. A series of 20 different, modern basecoat/clearcoat systems for automotive OEM were exposed outdoors on the aforementioned weathering site in Miami, Florida. After outdoor exposure, all specimens exhibited an almost identical intensity of mildew growth on their surfaces.

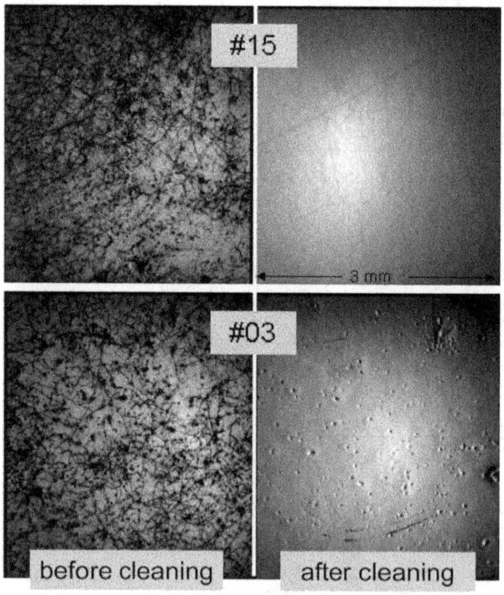

Figure 3.54: Surface pictures of two automotive coatings (#03 und #15) of different sensitivity to algal and mildew attack, after 14 weeks of outdoor weathering carried out in Miami, Florida (Okeechobee Road weathering site) in summer 2002,showing the state before (left column) and after cleaning (right column) of the weathered coating surfaces (from [49])

Magnifications of surface sections of the test panels of two different coating types are shown in Figure 3.54. The two pictures in the left column show the surfaces immediately after weathering and before cleaning. Differences in the intensity of mildew growth could not be recognized between coating #15 and coating #03. The two pictures in the right column show the surfaces after subsequent washing and polishing. In this state, clear differences were observed between the two coatings as regards the intensity of pinhole formation. Coating #15 does not exhibit any pinholes and therefore can be deemed resistant to weathering, including this special biological attack. In contrast, coating #03 shows poor weatherability. Since pinholes were formed to a rather high intensity, coating #03 could not be recommended for cars intended for use in regions with subtropical/tropical rainforest climate. The other coatings in the series ranked between the coatings #03 and #15 in terms of visibility of the surface damage. Figure 3.55 shows the relationship between mildew growth and coating damage for the example of coatings #03 and #06 from the same series. Both coatings were classified as having poor resistance to this special biological attack.

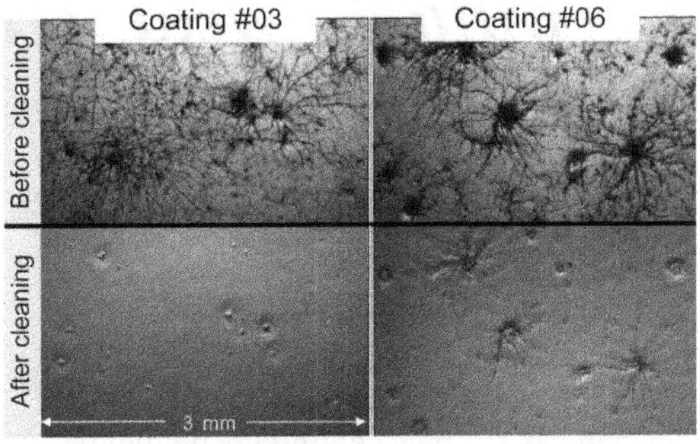

Figure 3.55: Surfaces of two automotive coatings sensitive to algal and mildew attack after 14 weeks of outdoor weathering in Miami, Florida (Okeeshobee Road weathering site) in summer 2002; the photographs show the state before (top) and after (bottom) cleaning of the weathered coating surfaces (from [49])

As in Figure 3.54, Figure 3.55 also shows the surface states before and after cleaning. A look at the photographs taken before (top) and after cleaning (bottom) reveals that pinholes always formed on the positions of most intensive mildew growth. Several repetitions of the outdoor weathering carried out at the same weathering site in the same season yielded the same weatherability rankings for the 12 coating samples.

Repetitions of the outdoor weathering at the same weathering site in other seasons, as well as at other weathering sites with subtropical/tropical rainforest climate produced little or marked formation of pinholes. Even outdoor weathering in Tully, Australia, in the eastern Australian rainforest and recommended as the most suitable for testing mildew resistance [50] failed to produce either more intensive mildew growth or more intensive formation of pinholes than in Miami, Florida. Additional studies proved that the presence of algae was necessary for the formation of pinholes. Furthermore, it was found that the mutual action of algae, mildew, global radiation, heat, and moisture was responsible for the formation of pinholes. However, only the special combination of weathering stresses that accidentally exists at the aforementioned weathering site in Miami, Florida, could have led to the extensive formation of pinholes there; this renders that weathering site the most suitable location for testing the resistance of automotive coatings to the type of damage that might arise in a tropical/subtropical rainforest climate.

However, the question remains: "What about the aforementioned special combination of weathering stresses acting at this weathering site in Miami, Florida?" Obviously, the worst case weather scenario for coatings in subtropical/tropical rainforest climate occurred here in a particularly high frequency and intensity. Furthermore, this example proves that a worst case weather scenario can also consist of a combination of several different, successive weather scenarios. In this case, two alternating types of weathering stresses can be presumed. First, the weather conditions at night (100 % R.H. at 25 °C) were optimal for growth of algae and fungi. Second, the weather conditions at noon (high UV irradiance and up to 60 °C) were optimal for the formation of pinholes caused by photodegradation that was probably boosted by organic substances produced, for instance, by algae or fungi at night.

It was observed that fungi mycelium grown during outdoor weathering never became so dense as to extensively prevent the global radiation from accessing the coating surface. Therefore, the aforementioned weathering site in Miami, Florida, could nevertheless be used for testing the weatherability of coatings that do not respond by forming pinholes. However, if the weathering-induced formation of pinholes on non-mildew-resistant coatings interferes with the assessment of weathering-related changes in gloss, colour, or other ageing criteria, it is possible to revert to the ATLAS NEW SITE, which is another weathering site near Miami, Florida, which opened in 2006 and has proved to be a location that does not undergo recognisable attack by mildew.

4 Outdoor weathering - natural form of accelerated testing

4.1 General aspects

"Have the coating tested by outdoor weathering!" may be the simplest answer to the creator's question: "Is my new coating good enough to resist all stress situations that occur in normal use outdoors?" Outdoor weathering is the natural form of accelerated testing. It uses the weathering stresses arising from the natural environment. This statement can evoke the questions: "Can outdoor weathering be regarded as accelerated testing at all? Can it fulfil the most essential requirements, such as time-compressed or accelerated coating ageing?" Without doubt, any outdoor weathering test which only employs normal use as the source of weathering stress cannot be regarded as accelerated testing. This was discussed in Chapter 2.4 for the example of certain studies by the automotive industry. As described there, mounting the test panels on the body of a car that is stressed by normal use can be seen as a special form of outdoor weathering - however, there is no time-compressing or accelerating effect, and therefore it is not accelerated testing.

Outdoor weathering performed at the site of normal use can only be regarded as accelerated testing if stressed, for instance, by increased radiant exposure. This is the case when, for instance, test panels of automotive coatings constantly face the equator, leading to exposure to more hours of sunshine than found on a frequently moved car.

As already mentioned in Chapter 2.5, service life prediction should also be possible in a shorter time if modern analytical methods are used to identify the coating's ageing behaviour much earlier than is possible through the use of common ageing criteria, such as gloss or colour. In this case, a time-compressed effect is attainable, too. Therefore, normal use combined with modern methods of analysis could also be regarded as a special version of accelerated testing. In certain cases, outdoor weathering without any time-compressed or accelerated effect may be necessary when conventional outdoor weathering leaves doubts about the accuracy of simulating a special kind of normal use.

Studies of the weatherability of aircraft coatings are a typical instance of the unavailability of suitable kinds of outdoor weathering performable on the ground or artificial weathering performable in the laboratory [51]. In this example, test panels were immobilised on the fuselage of a plane and subjected to the stress arising from normal long-distance flights. This version of outdoor weathering was free of uncertainties, in contrast to other versions being performed accelerated or time-compressed. However, high time consumption remained an essential disadvantage because the various worst case weather scenarios occurred with limited frequency. Furthermore, only a portion of the maximum possible global radiation that would exist at the site of normal use was used for testing. This is a problem for all products that are mostly in motion, such as street cars, trains, aircraft, and water vessels. Their position relative to the sun changes instantaneously. In some cases, direct irradiation is intentionally avoided (for instance, when cars are parked). However, immovable products also receive only a limited part of the available global radiation, for instance façade paints. The mostly vertical position of common house walls leads mostly to substantial deviation from the optimum exposure angle. Even the south-facing areas are affected. Furthermore, permanent or temporary shading by neighbouring buildings or vegetation is known to lower the receivable global radiation.

Ulrich Schulz: Accelerated Testing:
Nature and Artificial Weathering in the Coatings Industry

ISBN: 978-3-86630-908-1

Figure 4.1 shows a typical example of how a transparent, south-facing PVC building front suffered from uneven yellowing, obviously due to shady deciduous trees. The state of ageing after several years of use is shown in this picture. The photo was taken in winter, when the trees were bare. It can easily be seen that the areas behind the trees suffered less yellowing than the other areas which faced the sun unhindered. The areas behind the trees, which show up here as bright patches, suffered no extensive ageing. Obviously, the trees prevented the sun rays from reaching the façade's surface, especially during summer.

Figure 4.1: Uneven ageing of a building front made of PVC

Avoiding operational restrictions of natural weathering stress can be seen as offering the first potential for achieving a time-compressing effect. Let us assume in the following example that a car in normal use is exposed to only half the maximum available radiant exposure of global radiation. A conventional outdoor weathering test carried out in the same climatic region, but in optimum radiation conditions, would be expected to need only half the time to reach the same state of ageing. Of course, this presupposes a linear ratio between ageing behaviour and radiant exposure. Optimum radiation conditions afford the first essential potential for employing time-compression as a time-saving strategy. It can be reached by complying with the following restrictions: use test panels instead of finished or partly finished products, ensure that the test panels are of the optimum shape, the exposure angles are optimal, and avoid shaded areas.

The use of climatically favoured regions which provide greater weathering stress than encountered in the intended normal use offers the second essential potential for time-compression. Greater weathering stress leads to an expectation of faster coating ageing. According to the discussions in Chapter 2.5, those weathering sites are favoured for outdoor weathering of coatings whose respective worst case weather scenarios occur more frequently than in normal use, and whose respective weathering stress factors occur more frequently at their natural extremes. Outdoor weathering performed under such conditions fulfils all the requirements of a time-compressed test.

The worst case weather scenarios for newly developed coatings or coating systems are often not known. Therefore, suitable weathering sites became necessary that would allow at least the worst case weather scenarios listed in Table 2.4 to be realized. It was the coatings industry which started outdoor weathering, in the subtropical climate of South Florida, between 1920 and 1930 [52]. Suitable weathering sites were arranged by the paint industry as well as independent test institutions in accordance with the respective geographical situation and the availability of proper areas. The preference was to perform highly time-saving outdoor weathering in humid/hot, as well as dry/hot sites characterised by high values of annual radiant exposure and average ambient temperatures. Therefore, the search for suitable weathering sites was concentrated on regions between the northern and southern tropics (23°27' N and 23°27' S) or as close a possible to the equator. These are regions characterised by high angles of elevation to the sun. Worst case weather scenarios, as characterised by high irradiance combined with high values of temperature and extremely high or low values of relative humidity, can also be found during the summer month in regions with moderate climate. However, they occur much less often than in hotter climates.

Weathering sites in regions with cold or moderate climate do not offer substantial time-compression, mainly because of their lower annual average ambient temperature and their lower annual radiant exposure. Consequently, outdoor weathering in cold or moderate climates

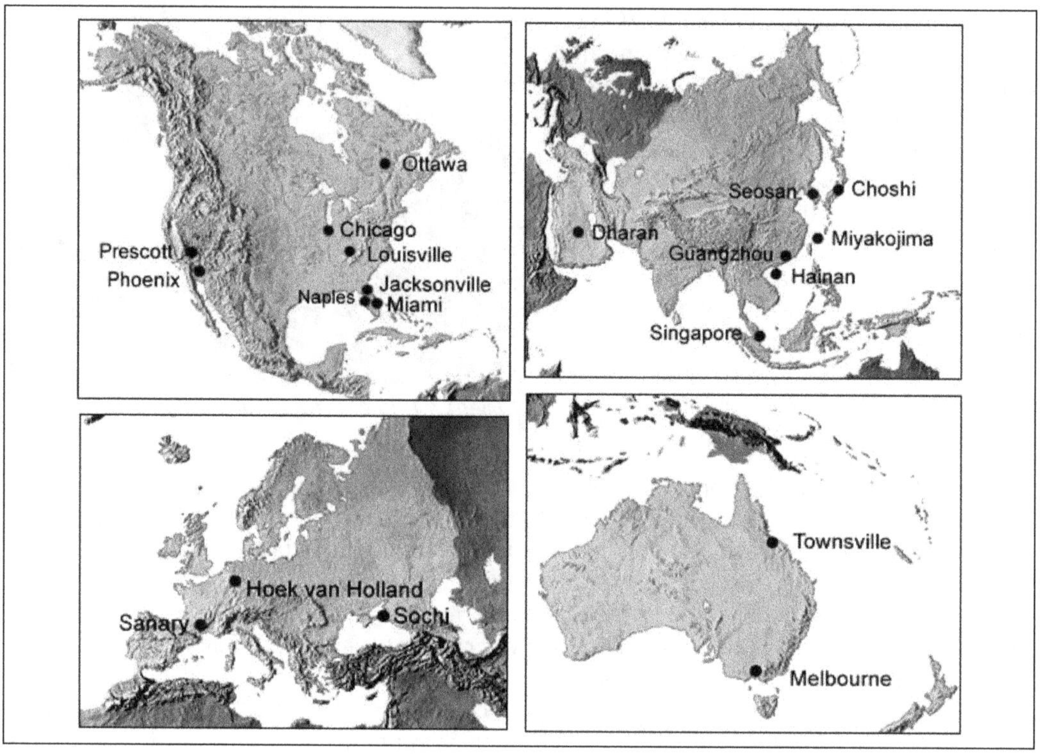

Figure 4.2: Important outdoor weathering sites operated by ASWG around the globe (adapted from [30])

has not gained major significance in coatings testing. Figure 4.2 shows by way of example an overall view of weathering sites run or operated by AWSG (Atlas Weathering Service Group), a privately run American company. The picture shows only a selection of the network of weathering sites distributed over the whole globe. As mentioned before, not all of these weathering sites have attained significance in coatings testing

The weathering sites located in South Florida, and the area near Phoenix, Arizona, have gained the most significance for coatings testing, and, therefore, are generally recommended for use as reference climates or reference stations for outdoor weathering.

4.2 Selection of the right weathering site

4.2.1 Outdoor weathering in reference climates

What is the point of a high acceleration factor for coating ageing if the outdoor weathering repeated in the following year yields completely different results? What is the point, for instance, if service life predictions based on the results of one year have to be corrected after outdoor weathering the next year? It is no rarity to observe that the weatherability rankings of coatings tested in more or less extensive coating series change, from year to year. There have been examples of paint makers having to change their opinion of what constituted their most weatherable coating when they repeated their outdoor weathering. Ranges that have been based on only one outdoor weathering test have had to be corrected.

However, in addition to high efficiency, each form of accelerated testing is required to deliver the most reliable and sure predictions possible of the service life that can be expected in normal use. The aforementioned examples led to a call for weathering sites which facilitate weathering stress that is repeatable and reproducible. Always using the same weathering sites is one way to expect increased repeatability of weathering stress. The other way is to find the weathering sites where the weathering stress is most repeatable from year to year. This in turn led to the call for weathering sites that could be justifiably declared reference climates or reference stations that could serve as standards.

Reference climates also proved necessary for making reliable comparisons of outdoor weathering tests carried out on different weathering sites. Furthermore, reference climates are needed for comparing outdoor weathering and artificial weathering. Reference climates must fulfil all the requirements generally imposed on reference standards. The main requirement is constancy of the standard's respective parameters or properties. In the case of reference climates, this means good repeatability of the weathering stress from test to test. However, good repeatability means not only repeatable climatological data, but also repeatable weathering results. As will be discussed later in Chapter 4.4, good repeatability of the essential climatological data need not lead to good repeatability of the weathering results. Of course, to be able to assess the repeatability of the weathering stress requires at least knowledge of the essential climatological data listed in Table 3.7.

As already shown in Figure 3.29, weathering sites operated in the moderate climate of central Europe cannot fulfil the requirements for a reference climate as there is too much variation in the global radiation. Due to the random character of the weather, repeatability is always limited at each weathering site. **Outdoor weathering always has a more or less limited repeatability that cannot be avoided and must be tolerated.** Therefore, a weathering site may only be declared a reference climate on the basis that repeatability factors in the natural limitations. However, a reference site can lose its status if other weathering sites prove to have better repeatability. The weathering sites in Miami, Florida, and Phoenix, Arizona, which are presently used as reference climates nonetheless exhibit remarkable annual variation in essential climatological data, such as global UV and rainfall. In Figure 4.3, the data for both reference climates are presented for the period January 2000 to December 2005. It cannot be avoided that the obvious variation in the data will inevitably lead to corresponding variations in the respective weathering results.

Looking at the variation in rainfall for Miami, Florida, we would be justified in distinguishing between wet years and less wet years (for instance, 2004 with 1178 mm and 2001 with 2032 mm).

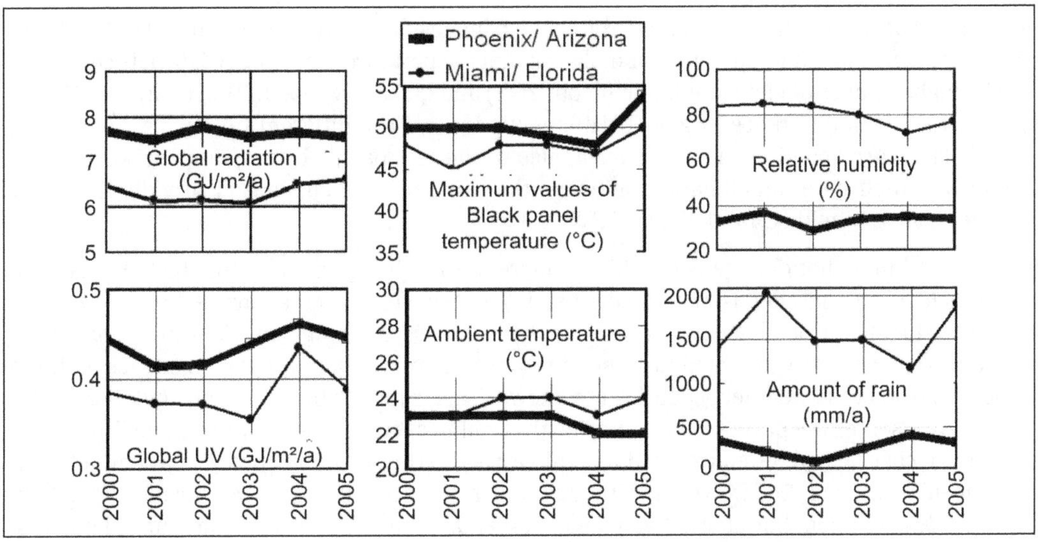

Figure 4.3: Annual sums and yearly average values of important climatological data for a 6-year-period from 2000 to 2005, published in [30], for the reference climates Miami, Florida, and Phoenix, Arizona

Table 4.1: Features of the climates of Miami, Florida, and Phoenix, Arizona, commonly used as reference climates (from [30])

North America **Operator: AWSG**	**Miami/Florida USA (26°N, 81°W)**	**Phoenix/Arizona USA (34°N, 112°W)**
Climate	sub-tropical humid	dry desert climate
Global radiation (GJ/m²/a)	6.6 (26° south)	8.4 (34° south)
	6.3 (5° south)	7.6 (5° south)
Global UV, 290 to 400 nm (GJ/m²/a)	0.38 (26° south)	0.43 (34° south)
	0.39 (5° south)	0.44 (5° south)
Elevation/altitude (m above MSL)	3	610
Distance from ocean (km)	27	ca. 500
Ambient temperature (annual mean) (°C)	23.4	22.8
Maximum black panel temperature (annual mean) (°C)	47.2 (26° south)	49.4 (34° south)
	45.4 (5° south)	45.0 (5° south)
Relative humidity (annual mean) (%)	81	34
Amount of rain (mm/a)	1521	263
Time of wetness TOW (%)	52 (5° south)*	2 (5° south)*

**other basis: 2000 und 2001*

Table 4.1 shows the essential climate data for the important North American reference climates, averaged over five years (2000 to 2004). Combined with high global radiation, these reference climates constitute the worst case weather scenarios most essential to coatings in general: the subtropical climate of Miami, Florida, typifies humid-hot weather scenarios (types 1 and 2 in Table 2.4), while the desert climate of Phoenix, Arizona, embodies the dry-hot weather scenario (type 3 in Table 2.4). The respective sites, operated by AWSG in Miami, Florida, und Phoenix, Arizona, are very important as regards test experience, test capacity, size of the testing area, diversity and throughput of samples. They are available to users all around the world. Both outdoor weathering sites are accredited test laboratories as defined in ISO 17025 [53]. This internationally respected accreditation is considered to offer proof of good test practice that is compliant and trustworthy.

Comparison of the aforementioned two reference climates fails to reveal any considerable differences between ambient temperature and black panel temperature. Phoenix, Arizona, is characterised by higher values of global radiation (about 20 %) and global UV (about 10 %) relative to Miami, Florida. The small difference in global UV is insignificant. The different tilt angles, annexed to certain data, correspond to the exposure angles of the test panels during outdoor weathering. In line with their different climatic characters, the two reference climates clearly differ in their average relative humidity, rainfall, and TOW.

Although, a humid-hot climate is considered more destructive than a dry-hot climate, it is common practice in paint chemistry to recommend that new coating creations be exposed to both reference climates. This is imperative as long as the coating's ageing behaviour in normal use is not known. However, exposure to just one of the two reference climates can be suggested if the respective worst case weather scenarios are known. This is mostly the case when outdoor weathering merely serves the purpose of monitoring the quality of current series production. Nowadays, automotive coatings are often exposed to outdoor weathering simultaneously in Miami, Florida, and Phoenix, Arizona, for two years. Although, two reference climates have proved adequate in past decades, additional or alternatively use of other weathering sites can sometimes be necessary or sensible. A good substitute for the Miami, Florida, reference climate is afforded by the other weathering sites in South Florida, such as those in the Everglades, Naples (both operated by

Table 4.2 Comparison of the reference climate Miami, Florida, with other weathering sites also located in subtropical tropical humid areas (adapted from [30])

Sub-tropic/tropic humid climates Operator: AWSG	Miami/FL USA (26°N; 81°W)*	Guangzhou China (23°N, 113°E)	Hainan China (19°N, 110°E)	Changi Singapore (01°N, 104°E)	Niyakojima (Okinawa) Japan (23°N, 113°E)	Townsville Australia (19°S, 147°E)*
Global radiation (GJ/m²/a)	6.6	4.6	4.7	6.0	4.9	7.2
Ambient temperature (annual mean) (°C)	23	22	24	27	23	25
Relative humidity (annual mean) (%)	81	79	81	84	76	70
Amount of rain (mm/a)	1521	1492	2013	2300	1741	937

**Accreditation according to ISO 17025*

AWSG) and Homestead (operated by Q-PANEL). These weathering sites are climatologically fairly identical with the reference site in Miami, Florida. It seems implausible that a coating which has proved adequately weatherable in Miami, Florida, would fail in a test performed on another South Florida test site. Nevertheless, be careful! There is always the possibility that other, unconsidered or unknown weathering stress factors, such as acid rain, aggressive air pollutants or biological attack will play an additional or predominant role in the process of coating ageing. In such cases, marked deviations from the expected weathering results cannot be ruled out. Furthermore, not all South Florida test sites are accredited to ISO 17025.

Additional use of weathering sites located in other parts of the globe is not expected to bring weathering results identical with those of the aforementioned two reference climates. This can only be recommended if the coatings are intended for worldwide use. In such cases, the relationship to the two reference climates would be of great interest. If this relationship is known, involving further weathering sites would be unnecessary.

The use of different weathering sites, whether only for the purpose of extending the testing area or whether as a competitor to an established reference site, causes problems with comparability of the weathering results. This assertion is supported by a comparison of the climatological data for the two reference climates Miami, Florida, and Phoenix, Arizona, with other sites which have the same climatic character, as shown in the following tables. Table 4.2 permits comparison with other sites which are all located in humid-hot regions, and Table 4.3, with another weathering site in a dry-hot desert climate. Only the weathering site in Townsville, Australia, is also an accredited test laboratory as per ISO 17025. Therefore, the other weathering sites described in this Table 3 cannot be expected to fulfil the strong quality regulations described in this standard.

Table 4.3 Comparison of the reference climate Phoenix, Arizona, with the weathering site Dhahran, Saudi-Arabia, both located in dry desert areas (adapted from [30])

Dry desert climate Operator: AWSG	Phoenix/ Arizona USA (34°N, 112°W)	Dhahran Saudi-Arabian (27°N, 50°E)
Global radiation (GJ/m²/a)	7.6	6.9
Ambient temperature (annual mean) (°C)	23	26
Relative humidity (annual mean) (%)	34	60
Amount of rain (mm/a)	263	80

As can be seen, the weathering sites classified under the same type of climate differ considerably in essential climatological data. The weathering sites with subtropical/tropical humid

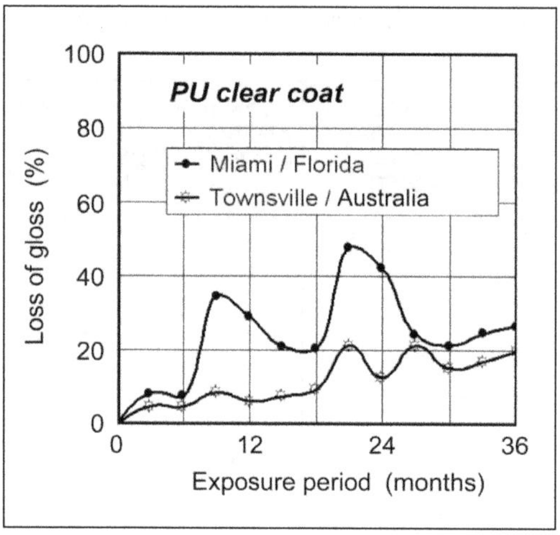

Figure 4.4: Loss of gloss by a white-coloured automotive coating, treated with a stabilised clearcoat during 3 years of outdoor weathering tests in Miami, Florida, and Townsville, Australia (black box, tilt 5 °)

climates differ markedly in global radiation (4.6 to 7.2 J/m²/a) and rainfall (937 to 2300 mm/a). The weathering sites with dry desert climates differ in their moisture conditions (34 to 60 % R.H.; 80 to 263 mm rain/a). These differences are greater than the yearly variation in data regularly measured in the respective reference climates. Therefore, changing from one reference climate to another weathering site cannot be recommended. In other words, **chosen reference climates should not be replaced by other weathering sites in spite of the same type of climate.** Repeating an outdoor weathering test in another weathering site will not lead to the expected degree of reproducibility.

Weatherability rankings for coating series tested at different weathering sites are mostly not identical. If the water stress differs substantially, comparisons of global radiation data cannot be used to assess the whole weathering stress. Calculations performed in this manner will yield false predictions. Figure 4.4 shows a typical example. Test panels of the same automotive coating were exposed outdoors in Miami, Florida, as well as in Townsville, Australia.

Both weathering sites are located in regions with a subtropical/tropical-humid climate. Comparison of the weathering results shows that the test panels exposed in Townsville lost less gloss in spite of 10 % higher annual radiant exposure (see Table 4.2). Clearly, the reason for the less pronounced ageing is the much lower rainfall. This reinforces the aforementioned recommendation that the same reference climates should always be used. Switches should be thoroughly thought over. They should only be done if an established reference weathering site has closed or another one with better weathering stress repeatability became available and could stake a justifiable claim to being classified as a reference climate. Long experience has shown that service life prediction based on weathering results from reference climates is commonly deemed to be reliable.

One last question remains to be answered: "How can the factors of time-compression or acceleration are calculated?" These factors are necessary in order that the weathering results from outdoor weathering in reference climates may be used to predict service lives for a certain type of normal use. For the sake of simplicity, the two factors are referred to as acceleration factor in the discussion below. The question may be answered with the following example. Let us assume that weathering results gained in Miami, Florida, are to be used to calculating the service life of coatings intended for normal use in central Europe. A rough estimation would proceed as follows. Based on the difference in annual radiant exposure, an acceleration factor of two, at least, can be assumed. Given that the average ambient temperature differs by about 10 °C, and the moisture conditions are almost similar, the acceleration factor can increase to at least four. We still have not allowed for the difference in the frequency with which the respective worst case weather scenarios occur. As already mentioned earlier, such calculations are characterised by high uncertainty. Nevertheless, they can be used for vague predictions.

Furthermore, theoretically calculated acceleration factors only take account of the increase in oxidation products (carbonyl groups, hydroperoxides, etc). The increase in oxidation products is often not proportional to the visually recognizable rate of ageing, as discussed in Chapter 3.2.2.3. Due to the mostly non-linear relationship between the increase in oxidation products and the change in common ageing criteria, such as gloss, haze, colour, scratch resistance, there may be a marked discrepancy between the calculated acceleration factors and those found in normal use. Furthermore, there is also a possibility that certain coatings tend to undergo faster ageing in normal use than in respective reference climates. This can be the case when other stress factors such as acid rain and biological attack occur in normal use that do not occur at the respective reference climates.

Acceleration factors based on long-term experience with certain types of coatings cannot be transferred without further ado to new products which have been developed, for instance, using new raw materials, formulations, or application methods. **Material-independent acceleration factors should not be used.** These are sometimes found in advertisements for certain weathering sites. Outdoor weathering performed over one or two years often leads to a measurable degree of coating damage that is still acceptable. In such cases, it is common practice to use these weathering results to interpolate for the expected service life. In other words, they are used to calculate the time until unacceptable damage occurs. Calculations of this kind are also very vague because coating damage seldom increases linearly with increasing exposure period. Such interpolations only make sense if the relevant relationships are backed up by experimental studies.

As described in Chapter 2.2, outdoor weathering is also used for corrosion tests on coated metallic substrates. The aim of such variants of outdoor weathering is to demonstrate the ability of the coating to protect the substrate against corrosion. To this end, tests in coastal regions are preferred. Corrosion of electrochemically negative metals is known to be promoted by the presence of salt aerosols, which are always to be found near the sea. Special weathering sites have therefore been established for this purpose in coastal regions. As discussed in detail in Chapter 2.2, such weathering sites have not been qualified as reference climates for outdoor weathering of coatings. Synergies arising from the interaction of global radiation and sea-salt aerosols are unknown. Worst case weather scenarios for modern coatings, including stress by sea salt, were not observed. Weathering sites that were used for corrosion tests failed to qualify as reference climates for outdoor weathering of coatings. However, the criteria for reference climates might be fulfilled by special European weathering sites restricted to corrosion tests, as described in Table 4.4, where the climatic conditions, especially global radiation, play a subordinate role relative to the sea-salt aerosols.

4.2.2 Outdoor weathering in European weathering sites

Some weathering sites have been established in Europe that conduct corrosion tests in which natural stress predominates. A selection of these is presented in Table 4.4 (page 72). As described before, coastal areas are preferred because the presence of sea-salt aerosols is necessary for accelerating the corrosion. European test sites have hardly attained any significance for outdoor weathering. Modern coatings that are sensitive to sea-salt aerosols are unknown. It should be remembered that sufficiently high temperatures are needed for global radiation to cause substantial destruction of coatings! In comparison to the climate of Miami, Florida, the European climate appears quite benign. European weathering sites therefore cannot become earnest competitors to the North American reference sites. The acceleration factors expected from the quite harmless European climate are unlikely to fulfil the criteria for accelerated testing. As can be seen in Table 4.4, the relative humidity of the atmosphere is similar to that of Miami, Florida, due to the sites' proximity to the sea.

However, due to the relatively low radiant exposure, low ambient temperature, and low rainfall, the European climate is barely likely to accelerate coating ageing relative to intended normal use. It is a waste of time to use European weathering sites for testing the weatherability of coatings. The long-established North American reference climates will always be preferred for coatings testing.

Table 4.4: Comparison of the reference climate of Miami, Florida, with European weathering sites located near the sea (from [30])

European sites (sea climates) Operator: AWSG	Miami/FL USA (26°N; 81°W)	Sanary/Bandol France (43°N, 6°E)	Hoek van Holland The Netherlands (52°N, 4°E)	Sochi Russia (43°N, 40°E)
Type of climate	sub-tropical humid	Mediterranean sea climate	moderate North Sea climate, with corrosive atmosphere	Black Sea climate
Global radiation ($GJ/m^2/a$)	6.6	5.5	3.8	5.0
Elevation/altitude (m above MSL)	3	110	6	30
Distance from sea (km)	27	approx. 4	<1	approx.1
Ambient temperature (annual mean) (°C)	23	13	10	14
Relative humidity (annual mean) (%)	81	76	87	77
Amount of rain (mm/a)	1521	700	800	1390

The weathering site located in the harbour area of Hoek van Holland, The Netherlands, is known for the additional stress posed by aggressive air pollutants from nearby industrial sources. This site is therefore still used sometimes in coatings testing to demonstrate the coatings' weatherability in such aggressive environments. Since global radiation and temperature boost the action of aggressive chemicals, outdoor weathering in the warmer and sunnier climate of Jacksonville, Florida, as described in Chapter 4.2.3, is commonly preferred, especially for testing automotive coatings.

4.2.3 Outdoor weathering in Jacksonville/Florida

As already described in Chapter 3.2.3, high climatic stress need not always represent the worst case weather scenario that boosts coating ageing. Anthropogenic influences, especially acid atmospheric precipitation, can be part of the weathering stress. Acid can generate a widespread type of surface damage, as shown in Figure 3.52. Starting in the early 1980s, at the same time as high-gloss automotive coatings found widespread use this type of surface damage was found first on imported cars that were temporarily stored outdoors in the harbour area of Jacksonville, Florida. The damage, which occurred after just a few months' exposure, was referred to as acid etching. When it was realized that acid atmospheric precipitation, in combination with the subtropical climate of Florida, was the reason for the coating damage, weathering sites were established among or near the places where the cars were stored. Since that time, weathering sites in Jacksonville, Florida, have been used to demonstrate the weatherability of automotive coatings in regions where acid atmospheric precipitation is generally a part of the weathering stress. Situated in northern Florida, the Jacksonville harbour area is characterised by its proximity to several oil or coal fired power stations and air-polluting industrial plants. This special environmental situation may be clearly seen in Figure 4.5, which is a photo taken from a weathering station in Jacksonville. Combined with humid subtropical climate and the typical atmosphere of an overseas port, Jacksonville, Florida, appears to be the best available weathering site for this special purpose. It is currently mostly used for testing automotive coatings.

Table 4.5 compares the essential climatological data from Jacksonville, Florida, with those from the Miami, Florida, reference climate. With its more northerly location, Jacksonville experiences lower global radiation and a lower ambient temperature. Due to the lower relative humidity and lower rainfall, the climate of Jacksonville can be considered to be less damp than that of Miami. Unfortunately, regularly measured data for acid stress are not available. This is a major disadvantage, because acid stress is known to vary considerably from year to year. Experience has shown that acid etching is more prevalent during the warmer and more humid summer months from May to September. This is evidence that the destructive effect of acid atmospheric precipitation can only fully unfold in humid-hot climates. Since the other months do not contribute significantly to the extent and visibility of acid etching [45], it is common to expose automotive coatings only during the summer time. Since 14 weeks have proved to be long enough, exposure commonly starts in the first week of June and ends in the second week of September.

Figure 4.5: Photo of a weathering site in Jacksonville, Florida

Table 4.5: Comparison of the outdoor-air climates of the Florida weathering sites located in Jacksonville and Miami (adapted from [30])

Florida/USA Operator: AWSG	Jacksonville (30°N, 82°W)	Miami (26°N, 81°W)
Climate	sub-tropic & acid and sea salt containing precipitation	sub-tropic
Global radiation (GJ/m²/a)	5.8	6.3
Elevation/altitude (m above MSL)	8	3
Distance from ocean (km)	15	ca. 30
Ambient temperature (annual mean) (°C)	20	23
Relative humidity (annual mean) (%)	76	81
Amount of rain (mm/a)	1303	1521

Unfortunately, the stringent criteria for accreditation to ISO 17025 cannot be fulfilled by the Jacksonville weathering stations. Therefore, these weathering sites cannot be recommended without further ado as reference climates. First of all, reliable data for acid stress are lacking. It is not possible to compare the acid stress from one year to the next. Measurements of precipitated acid are not easy to take. Simultaneous depositions of wind-borne limestone particles from the coral soil render collection of acid problematic. However, knowledge of all weathering stresses is necessary for the assessment of repeatability. Therefore, it is essential that a corresponding set of reference coatings be exposed simultaneously at each outdoor weathering site in Jacksonville, Florida. That is the only way to assess repeatability of the weathering results well. As will be discussed in more detail later in Chapter 4.4, sensibly chosen coatings of known weatherability that has been demonstrated in previously performed outdoor weathering in Jacksonville, Florida, should be used as reference coatings for this purpose.

Until now, the owners of the coatings to be tested have also provided the reference coatings. Only by means of simultaneously exposed reference coatings is it possible to decide whether or not the results of a certain outdoor weathering test can be included in predicting the service life. Sometimes, the staff involved in coatings testing experienced years when no acid etching occurred on their test panels. In this case, stored weathered reference samples

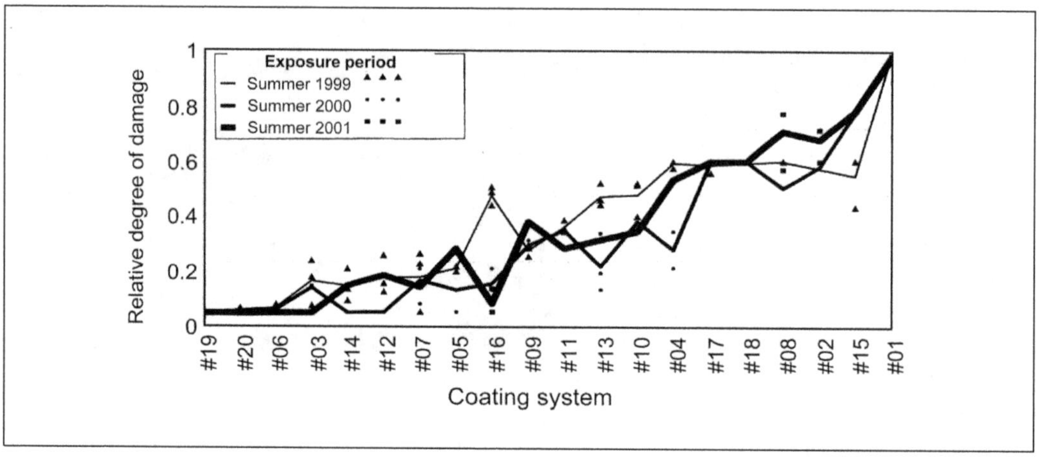

Figure 4.6: Annual fluctuation in the visually assessed degree of damage to identical series of 20 different automotive coatings weathered outdoors in Jacksonville, Florida, for 14 weeks during three different summers (from [45])

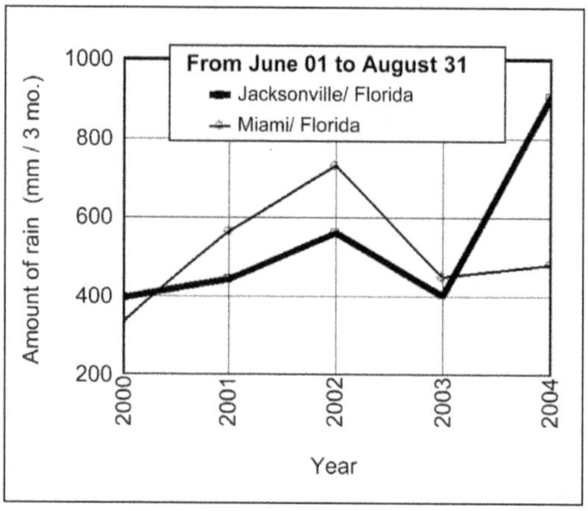

Figure 4.7: Annual fluctuation in rainfall published by [30] for Jacksonville und Miami for the period from 1st June to 31st August from 2000 to 2004

displaying the highest coating damage ever caused by outdoor weathering in Jacksonville, Florida, would serve as a very important tool for deciding whether the test must be repeated or not. Figure 4.6 illustrates the limited repeatability of the weathering results for the example of a series of 20 different basecoat/clearcoat systems for automotive OEM. Three outdoor weathering tests carried out on identical coating series in the summers of 1999, 2000, and 2001 are compared. In Figure 4.6, the visually assessed degree of damage in the different years is plotted against the respective coating numbers (#01 to #20). The coating systems were arranged in increasing order of damage.

As can be seen, the coatings respond to different extents to the annual variations in weathering stress. For instance, coating system #15 exhibited its best weatherability in 1999, while coating system #04 and coating system #16 performed best in 2000 and 2001 respectively. Whether the variations in acid stress or the variations in climatic stress factors, such as global radiation, heat, or moisture, are responsible for the fluctuation in the weathering results cannot be answered. Figure 4.7 shows, that fluctuations in rainfall could be a reason. The fluctuation in the Jacksonville data is similar to that for Miami, and also allows more or less damp years to be distinguished.

Despite not having ISO 17025 accreditation, Jacksonville, Florida, is the only commonly available weathering site that can currently be recommended for testing the weatherability of coatings intended for use in all regions where acid atmospheric precipitation is part of the weathering stress. Furthermore, the test needs only a relatively short time of 14 weeks. Since only the summer season can be used for the test, outdoor weathering in Jacksonville, Florida, logically be performed only once a year. Acid-stressed weathering sites in cooler regions cannot be recommended as they are disadvantaged by considerably lower test effectiveness.

4.3 Conventional and intensified outdoor weathering

4.3.1 Conventional outdoor weathering (constant exposure angle)

Outdoor weathering carried out at constant exposure angle and without any human interference in the coating surface micro climate of the test panels (open back) is the form most often used and is therefore referred to as conventional outdoor weathering in this book.

It is characterised by exposing the test panels at a constant exposure angle and direction during the whole test. Selecting the most appropriate exposure angle has always been one of the common problems of outdoor weathering test design. As already discussed, the direct portion of global radiation is best used for testing if the test panels are oriented towards the equator at an exposure angle corresponding to the test location's degree of latitude. According to the Beer-Lambert law, vertical incidence of the sun's rays leads to highest irradiance. In other words, under a clear sky, the test panels will be subjected to the highest irradiation stress if directed towards the sun at noon when the direct portion of global radiation reaches its maximum irradiance. When the direct portion only is considered, an exposure angle in accordance with the latitude was always seen as optimal. For the two reference climates, exposure angles of 26° for Miami, Florida, and 34° for Phoenix, Arizona, are common. An exposure angle of 45° is preferred in central Europe. Since this statement is only valid for the direct portion, as discussed in more detail in Chapter 3.2.1.3, other exposure angles may be optimal for regions with higher diffuse solar radiation. Other exposure angles have also been used for certain applications. Figure 4.8 shows a section of the test area of a weathering site located in South Florida, where the test panels were exposed at two different angles.

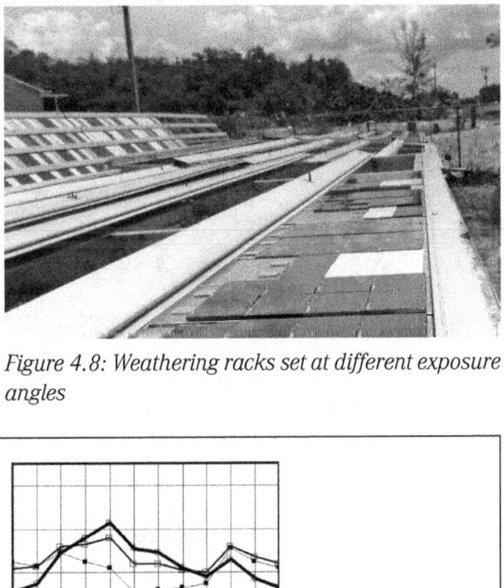

Figure 4.8: Weathering racks set at different exposure angles

According to the intended application of the coated products or where special studies are carried out, the exposure angles range from 0° (horizontal) to 90° (vertical) facing south, north or other directions. 5° facing south is preferred for coatings testing if the horizontal orientation of the coated surface is prevalent in normal use. That is especially the case for roofs

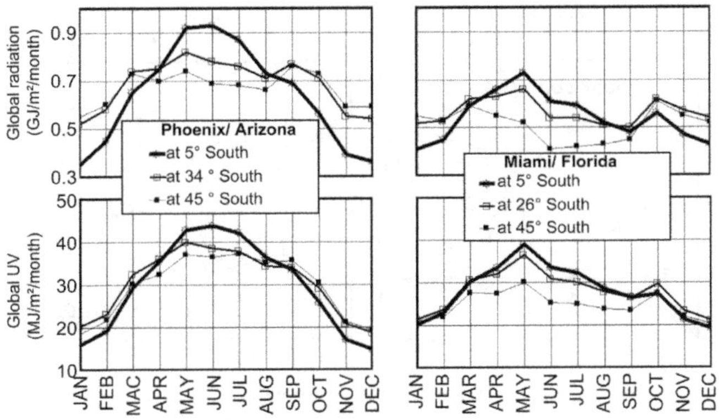

Figure 4.9: Monthly radiant exposure (global radiation in the top series, global UV in the bottom series) published by[30] for the reference climates Phoenix, Arizona (left) and Miami, Florida (right) as a function of months and exposure angle during 2004

of automobiles or special flat roofs of buildings. Outdoor weathering in Jacksonville, Florida, commonly prefers 0° in order to prevent precipitated acid rain from running off the coating surface. As there are so many variants, test engineers often raise the question: "Which of the offered exposure angles is the best one for my purpose?" First let us answer this question by comparing the effects of the most common south-facing angles for the example of the two North American reference climates Miami, Florida, and Phoenix, Arizona. Figure 4.9 compares global radiation and global UV for data from 2004.

The seasonal fluctuation in the elevation angles of the sun is reflected in the curves. That is particularly recognizable in the diagram for Phoenix, Arizona. Global radiation and global UV were almost identical. The two larger exposure angles (34° and 45°) returned roughly the same results. The curves for 5° and 45° differ considerably. As can be seen, the larger exposure angle yielded higher monthly radiant exposure than the smaller one did during wintertime. The summer season produced opposite results. The crater-like shape of the curves for Miami, Florida, was obviously caused by the more intensive clouds during the summer months.

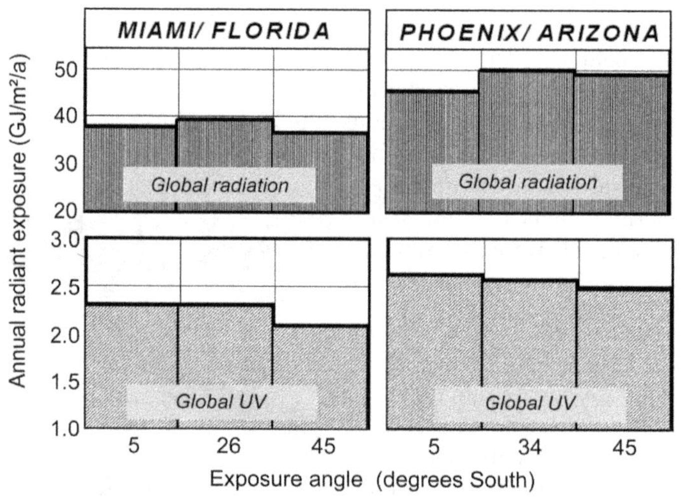

Figure 4.10: Mean annual radiant exposure (global radiation in the top series and global UV in the bottom series) published by [30] for the reference climates Phoenix, Arizona (left) and Miami, Florida (right), averaged over five years as a function of the exposure angle

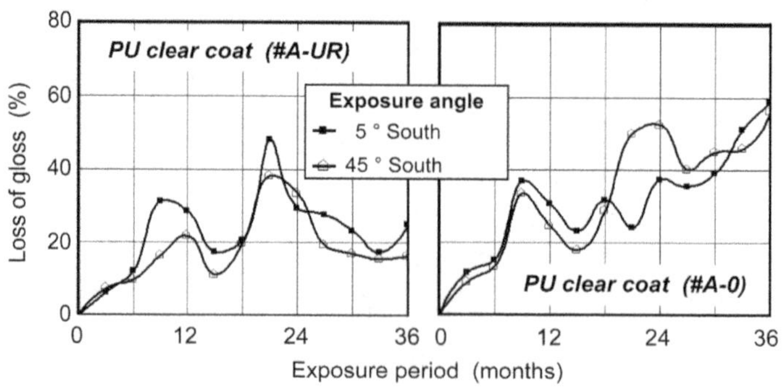

Figure 4.11: Loss of gloss of two differently stabilised clearcoats (#A-UR: stabilised, #A-0: unstabilised) on white-coloured basecoat, conventionally weathered outdoors in Miami, Florida, as a function of exposure period and exposure angle (adapted from [22])

Figure 4.10 shows the annual radiant exposure for global radiation and global UV as a function of exposure angle for the same reference climates. In contrast to the monthly totals, the annual totals show only very small differences of less than 10 %. From this result, it may be inferred that the commonly used south-facing exposure angles cannot be expected to produce differences in radiation-induced coating ageing, provided that outdoor weathering is not performed for less than one year, as commonly recommended.

So, with regard to annual irradiation stress, it does not matter as to which of the commonly offered exposure angles were used to weather the test panels outdoors in Miami, Florida, or Phoenix, Arizona. This assertion is also reflected in the weathering results, as the following example reveals. Specimens of two differently stabilised automotive coatings were exposed outdoors in Miami, Florida, at two different exposure angles. The loss of gloss over three years is shown in Figure 4.11. Aside from the marked temporal variation in the curves, there is significant dependence on the exposure angle.

Now, let us look at the angle-related effects in terms of the VIS and NIR parts of the global radiation. VIS and NIR are known to contribute only weakly to the photooxidation of common coatings. Since VIS and NIR are less scattered on their journey through the atmosphere, the test panels will always be subjected to highest thermal stress if the sun rays fall vertically on the surface. A substantial influence on the coating surface micro climate was expected. However, the values of BTT_{max}, published by [30], varied only slightly as a function of the exposure angle. The monthly average values differ only by 6 K and the annual average values only by 4 K. **So, it is also of no consequence for the surface temperature whether the test panels are exposed at 5, 26, 34, or 45° towards the equator.**

Figure 4.12: Black and white test panels mounted on a vertical weathering rack (exposure angle: 90°)

What about the influence of the exposure angle on the moisture conditions on the coating surface micro climate? Let us look at the TOW. Considerable angle-dependence was found for the years 2000 and 2001 in Miami, Florida, again published by [30]. Differences of up to 1390 hours, corresponding to 30%, were measured. The test panels exposed at the smallest angle (5°) revealed the longest TOW, clearly due to more intensive formation of dew and the fact that it runs off more slowly, as already described in Chapter 3.2.5.

An exposure angle of 90° (vertical) is used for outdoor weathering where the coatings are used for vertical parts, for instance, walls of buildings, window frames, and doors. Figure 4.12 shows a typical weathering rack of vertical black and white test panels.

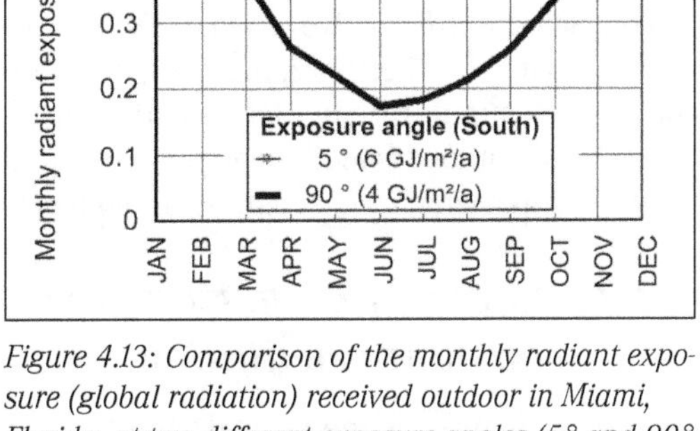

Figure 4.13: Comparison of the monthly radiant exposure (global radiation) received outdoor in Miami, Florida, at two different exposure angles (5° and 90° south respectively) as a function of month - in brackets: yearly totals (adapted from [30])

Figure 4.13 shows the influence of the exposure angle (south-facing) on the monthly radiant exposure received by the test panels. The

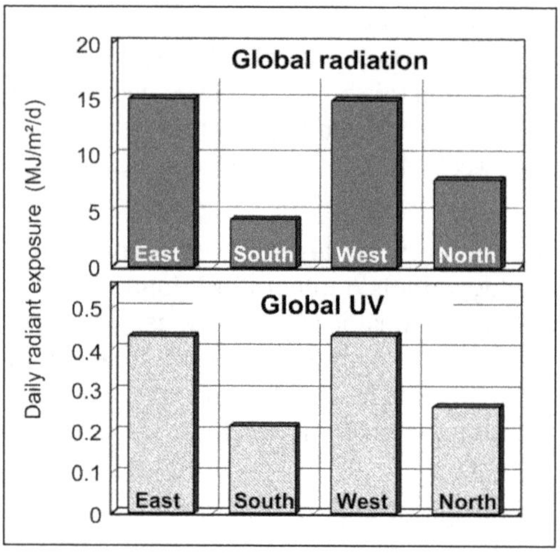

Figure 4.14: Daily radiant exposure (global radiation and global UV) on vertical test panels facing different directions, calculated for a cloudless summer's day in Miami, Florida (using CESORA [28])

data for 5° and 90° for one year in Miami, Florida, are shown. As may be seen, the monthly change in the sun's angle of elevation clearly influences the monthly radiant exposure. The biggest difference occurs during summer when the sun reaches its highest angle of elevation. Comparison of the yearly totals (in brackets) shows that vertical test panels receive about 60% of the radiant exposure that they would receive in the almost horizontal position.

The weathering racks normally face the sun for maximum solar effect. Exposure angles facing away from the sun are generally used when this simulation is more appropriate, or when exposure to direct solar radiation hampers the specific weathering effect being studied. Studies of special façade paints proved that facing south need not always be the most effective way to determine coating ageing. Test panels facing away from the sun do not attain a surface temperature higher than that of the ambient atmosphere as the direct portion of the global radiation is completely lacking. Therefore, water deposited by dew or rain stays on the panel surface longer.

The relative humidity of the coating surface micro climate is almost identical with that of the open air. Although test panels facing away from equator do not receive any direct radiation, a large part of the diffuse portion can arrive at the test panel surface. It comes from that part of the sky which is faced by the test panels, as discussed in more detail in Chapter 3.2.1.3. As is known, scattering is the cause of diffuse radiation. Since scattering increases with decreasing wavelength, sufficient UV radiation reaches the panel surface in the diffuse portion of the global radiation, and is able to cause marked coating ageing. As discussed in Chapter 3.2.5, there is a possibility that facing north proves to be the worst case weather scenario for certain coatings, due to the interaction of high moisture content, high ambient temperature and global UV, as in Miami, Florida.

Figure 4.14 characterises the irradiation conditions on the test panel surface arising from vertical exposure to the four compass points. The respective data for the global radiation as well as global UV were calculated for a cloudless June day in Miami, Florida, by means of CESORA. Due to sun's the high angles of elevation in June, the test panel facing east and west are subjected to the maximum, almost identical daily radiant exposure. Surprisingly, test panels facing north; receive considerably more global radiation than those facing south, even though no direct radiation reaches the panel surface. Further calculations for the same June day showed that the test panels facing north receive at least 25% of the global radiation that they would receive on a horizontal surface. Furthermore, more algal and fungal growth was observed on test panels which do not face the sun during outdoor weathering. According to the discussions in Chapter 3.2.7, it is possible that outdoor weathering designed in this way meets the worst case weather scenario for those coatings which are sensitive to biological attack.

4.3.2 Outdoor weathering at variable exposure angles and constant orientation

As seen in Figure 4.9, the ratio of the values for monthly radiant exposure measured at different exposure angles varied from month to month throughout the year. The one exposure angle brought the highest radiant exposure in the summer, and the other one in the winter. This led to proposals to change the exposure angle in several discrete steps with a view to enhancing annual radiant exposure. In one case, it was recommended that the weathering racks be readjusted four times a year in order that the exposure angle would track the change in azimuth angle throughout the year. The southwards orientation was to be retained.

Easy readjustment is a common option on modern weathering racks. The operator of the weathering sites offered respective proposals for Miami, Florida, and Phoenix, Arizona, as described in Table 4.6. This table lists the exposure angle which needs to be adjusted at set periods during the year. The data for global radiation and global UV for 2004, published by [30] were used in an attempt to prove whether the attainable effect justifies this regular intervention in the normal course of outdoor weathering and the additional effort of readjusting the weathering racks. The respective discussions were also thought to serve as an exercise in the use of published data for calculating the irradiation stress.

In accordance with the aforementioned proposal for Miami, Florida, the irradiation data are listed in Table 4.7 for each month, along with the belonging exposure angles. The line below December shows the totals for the monthly data, i.e. the annual figures for this method. The other three lines show the annual data reached by conventional outdoor weathering at constant exposure angles (5°, 26°, or 45° facing south). Comparison of the annual data in the last four lines would suggest that the proposal should be ignored!

Table 4.6: Programmes for outdoor weathering involving several discretely changing exposure angles (south-facing) designed for the reference climates Miami, Florida, und Phoenix, Arizona (adapted from [30])

	Exposure angle (latitude tilt angle) degree south	
Exposure period	Miami/Florida	Phoenix/ Arizona
1st Sept. till 31st Oct.	26	34
1st Nov. till 28th Feb.	45	54
1st Mar. till 30th Apr.	26	34
1st May till 31st Aug.	5	14

Table 4.7: Calculation of the increase in radiant exposure (global radiation and global UV) attained by the test method described in Table 4.6 versus that attained by conventional outdoor weathering at a constant exposure angle (5°, 26°, or 45° facing south) at the Miami weathering site, based on the irradiation data published by [30] for 2004

Miami/ Florida *month*	**Exposure angle degree south**	**Global radiation (300 to 2500 nm) *MJ/m²***	**Global UV (300 to 400 nm) *MJ/m²***
Jan	45	554	27
Feb	45	533	29
Mar	26	624	40
Apr	26	635	42
May	5	726	51
Jun	5	606	43
Jul	5	590	42
Aug	5	524	37
Sep	26	496	34
Oct	26	618	39
Nov	45	551	29
Dec	45	525	26
Jan till Dec	variable	6981 (100 %)	439 (100 %)
	5° constant	6491 (93 %)	435 (99 %)
	26° constant	6769 (97 %)	434 (99 %)
	45° constant	6129 (88 %)	385 (88 %)

4.3.3 Outdoor weathering on racks that continually track the sun

As mentioned before, the angle of incidence of the sun's rays and the sun's angle of elevation vary continuously from sunrise to sunset and during the year from 1st January to 31st December. As calculated for Phoenix, Arizona, the angles of incidence can vary by up to 80° during the day and up to 45° from noon to noon in the course of the year. According to the pertinent optics laws, continuous adjustment of the test panels to maintain vertical incidence of the sun's rays has always been considered to offer a way to increase the radiant exposure, at least, for the direct portion of global radiation. This can be achieved by means of a tracking rack. Mostly motorized, this machine automatically follows the sun's diurnal path from sunrise to sunset, ensuring that the sun's rays always meet the test panel surface vertically.

Modern tracking racks are automatically adjusted to the maximum quantity of direct radiation by means of measuring instruments, e.g. the pyrheliometers described in Chapter 3.2.1.6. Clearly, such weathering machines can reasonably only be set up in regions where clear skies are prevalent, e.g. on weathering sites located in desert areas like Phoenix, Arizona. First, let us answer the question: "What intensifying effect can be gained by means of this special type of outdoor weathering relative to conventional outdoor weathering?" The reference climate of Phoenix, Arizona, was chosen as prime example for answering this. The calculation was performed by CESORA and was based on the Beer-Lambert cosine law.

The results of the calculations are presented in Figure 4.15. The diagram on the left side shows, first, the change in the sun's angle of elevation during a clear 15th June and, second, the change in the cosine factor during the same day. The diagram on the right side shows, first, the sun's angle of elevation at noon for each 15th day of the month and, second, the monthly average of the cosine factor. This factor can be viewed as a measure of the increase in the radiant exposure that can be gained by permanent vertical incidence of the sun's rays. Since outdoor weathering is usually performed for a full year, at least, only the yearly data are of interest. Averaging for the whole year reveals a cosine factor of about 1.5.

Since the calculation was done under the restriction of a clear sky, the cosine factor calculated in this way describes the maximum achievable intensification. **Annual radiant exposure therefore can be enhanced relative to conventional outdoor weathering by a factor of 1.5 at most through continuous weathering on racks that track the sun.** Figure 4.16 shows a photograph of a typical tracking rack that permits dual axis tracking during outdoor weathering. Note that the calculation only considered the direct portion of the global radiation! The diffuse portion, which also occurs in a clear sky, can also be expected to influence the intensification factors. However, both the direct and the diffuse portion contribute to the irradiation stress acting on test panels mounted on the tracking rack.

The following discussions are therefore devoted to the question as whether diffuse radiation contributes substantially to the intensifying effect. Figure 4.17 is intended to serve as a tool for making a rough estimate.

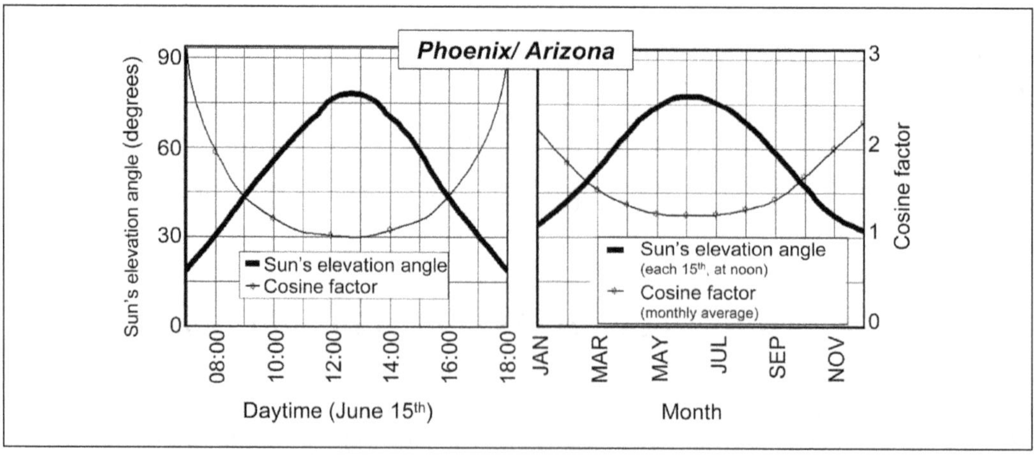

Figure 4.15: Sun's angle of elevation and the cosine factor during a clear 15th June (left diagram) and during a whole year (right diagram), calculated for Phoenix, Arizona, by means of CESORA [28]

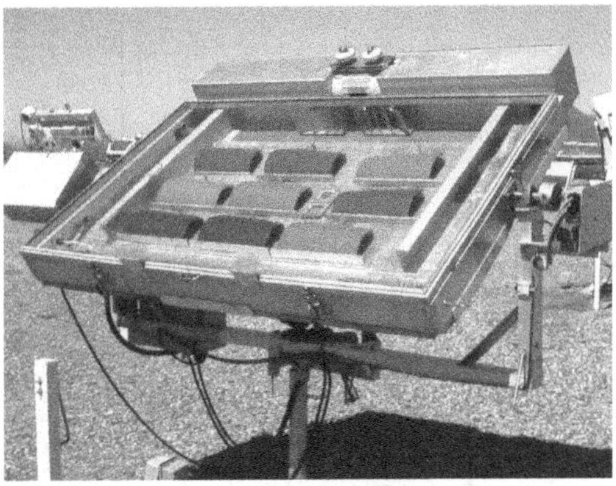

Figure 4.16: Typical dual-axis tracking rack for outdoor weathering (adapted from [30])

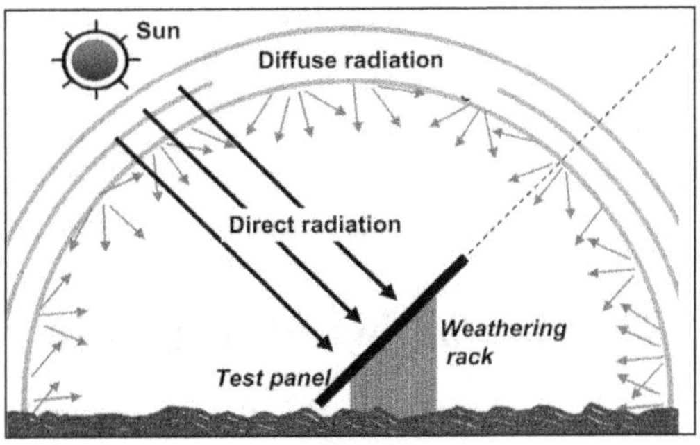

Figure 4.17: Sketch for illustrating and roughly estimating the intensifying effect of dual axis tracking of the suns' path.

The situation at 4 p.m. local time on a mid-June day in Phoenix, Arizona, was chosen as a good example for these discussions. In this situation, the sun is about 45° above the horizon. The tracking rack should have adjusted the test panel automatically to an exposure angle of 45° to ensure that the sun's rays (thick black arrows) fall vertically on the test panel surface. In this position, the test panel is stressed by the maximum possible irradiance at this moment due to direct radiation. According to the cosine of the sun's angle of elevation, the irradiation intensification factor relative to a horizontal test panel reaches about 1.5. The diffuse radiation comes from every direction in the sky. The sketch shows that the test panel surface faces only a part of the sky - less than half - at an exposure angle of 45°. In other words, only a portion of the maximum possible available diffuse radiation meets the test panel surface.

Let us stay with this example to discuss the intensification effect of dual-axis tracking when both portions of the global radiation act together. The small wavelength range of $(320 \pm 0.5)\,\mu m$ was intentionally chosen for these discussions. This wavelength range is known to be the most active in promoting coating ageing. From Figure 3.24, it can be seen that 40 % of the solar radiation arrives at the test panels in direct form and 60 % in diffuse form, at 320 µm. In other words, only 40 % of the solar radiation can be intensified by adjusting the test panels to achieve vertical incidence of the sun's rays, according to the Beer-Lambert cosine law. As for the diffuse portion, some of is prevented from impinging on the test panel surface. Since the area of sky facing the test panel is reduced due to adjustment to a 45° exposure angle, only about 75% of the maximum available diffuse solar radiation impinges on the test panel surface.

In contrast to the case for direct radiation, tracking diminishes the contribution of diffuse radiation. In this example, intensification of direct radiation and diminution of diffuse radiation occurred to about the same extent. This example shows that **the positive effect of direct radiation and the negative effect of diffuse radiation can cancel each other out where outdoor weathering is performed on tracking racks. As for the most effective radiation at 320 nm, weathering on the tracking rack failed to reveal higher irradiation stress than that for conventional horizontal outdoor weathering.**

A further calculation based on all the global UV from 300 to 400 nm was performed on the annual data published by [54], and on the reference climate for Phoenix, Arizona; this revealed only an insignificantly better intensification factor of 1.2.

Tracking would be expected to increase the temperature of the coating surface. However, as discussed in Chapter 3.2.5, an increase in the temperature leads to a decrease in the relative humidity of the coating surface micro climate. Therefore, an acceleration of coating ageing by tracking due to a higher test panel temperature cannot be expected. Tracking can reasonably only be used at weathering sites characterised by a prevalence of clear skies, as typically occur in dry desert climates. The Miami reference climate with its high clouds, which occur especially often during

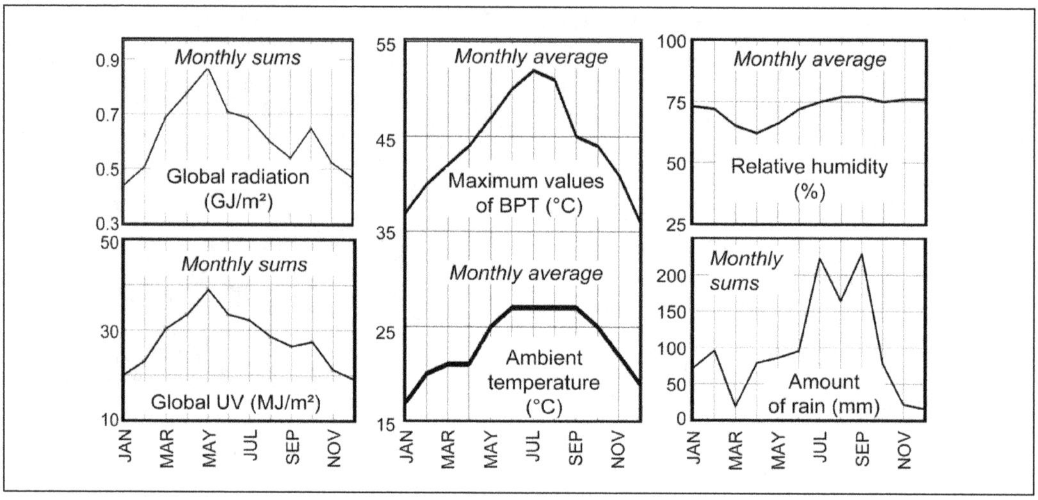

Figure 4.18: The parameters of the reference climate of Miami, Florida, published by [30] for 2004

the most effective summer months, is unsuitable for this kind of intensified outdoor weathering. Since humid-hot climates prove to have a greater effect on coating ageing, tracking has not gained appreciable significance in coatings testing.

4.3.4 Outdoor weathering with change in weathering site

As shown with the example of the reference climate Miami, Florida, for 2004 in Figure 4.18, all climatological data always display a clear maximum during the summer season over the course of the year.

Apparently, this observation led to a new proposal for achieving time-compression. The idea was that the annual radiant exposure would be enhanced by combining the Miami reference climate with a similar climate found on the southern half of the globe. The aim was to achieve a two-summer exposure in one 12-month period by changing the weathering site once a year. The subtropical/tropical-humid summer climate proved to be more effective at ageing coatings than that of the other seasons. The operator of a weathering site in Miami, Florida, offers a special weathering schedule that it calls the EverSummer Programme. This weathering programme suggests the weathering site in Townsville, Australia (19° south), as described in Table 4.2, as being a suitable counterpart for the Miami, Florida, test site (26° north). Figure 4.19 compares two of the most important climatological data published by [30] and [50] for Miami, Florida, and Townsville, Australia, for 2003. As expected, the monthly radiant exposure (global radiation) and the total monthly rainfall are the opposite of each other during the course of the year.

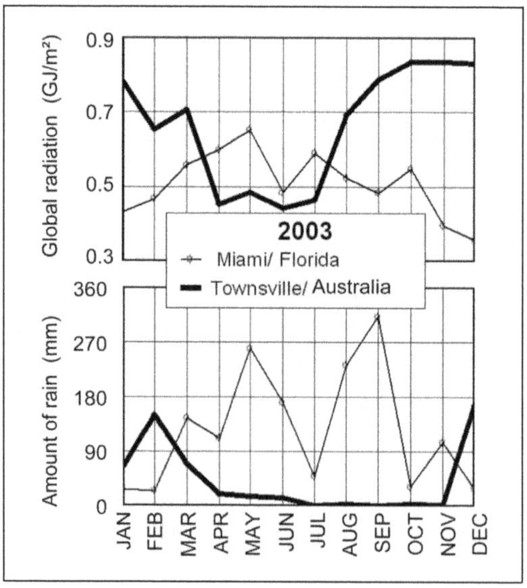

Figure 4.19: Change in the monthly totals of global radiation and rainfall in the course of 2003 for the weathering sites Miami, Florida, and Townsville, Australia (adapted from [30 & 50])

The EverSummer Programme suggests exposing the test panels from 21st March to 20th September in Miami, Florida, and then directly from 21st September to 20th March of the following year in Townsville, Australia, the supposition being that the test panels can be shifted quickly from one location to the other. Unfortunately, there are as yet no published weathering results that confirm a time-compressing effect on coating ageing. Therefore, an attempt has been made to calculate at least the intensification factor using the published climatological data for the aforementioned 12-month period. The results for the EverSummer Programme are presented in the fourth column of Table 4.8, opposite those for a simultaneously performed, uninterrupted exposure in Miami, Florida (column 3).

The differences between these two types of outdoor weathering are shown in columns 5 and 6. The EverSummer Programme seems able to intensify outdoor weathering by a factor of 1.3 (28 %) for global radiation and 1.5 (46 %) for global UV. No difference is found for the mean ambient temperature. However, relative humidity and rainfall reveal negative effects. Figure 4.20 com-

Table 4.8: The most important weathering parameters attained by a conventional outdoor weathering in Miami, Florida (column 3) in comparison to those attained with the EverSummer Programme (column 4), along with the advantages or disadvantages (column 5), based on data published by [30, 50] for the period 21st March 2002 to 20th March 2003

EverSummer **Weathering parameters**		**1 year Miami** **21st March 2002 till 20th March 2003**	**1 year EverSummer** **Miami*: 21st March 2002 till 20th Sept. 2002 & Townsville**: 21st Sept. 2002 till 20th March 2003**	**Advantages (+) and disadvantages (-) (differences: 4 minus 3)** **Absolute values**	**(Relative values)**
1	**2**	**3**	**4**	**5**	**6**
Ambient temperature (annual mean)	°C	24	24	0	(0 %)
Relative humidity (annual mean)	%	84	75	-9	(-11 %)
Global radiation (300 to 2500)	GJ/m²	6.2	8.0	1.8	(+28 %)
Global UV (300 to 400)	GJ/m²	0.38	0.55	0.18	(+46 %)
Amount of rain	mm	1531	1339	192	(-13 %)

* Exposure angle: 5 ° South (adapted from: www.atlaswsg.com)) ** Exposure angle: 0 ° (adapted from: www.allunga.com)

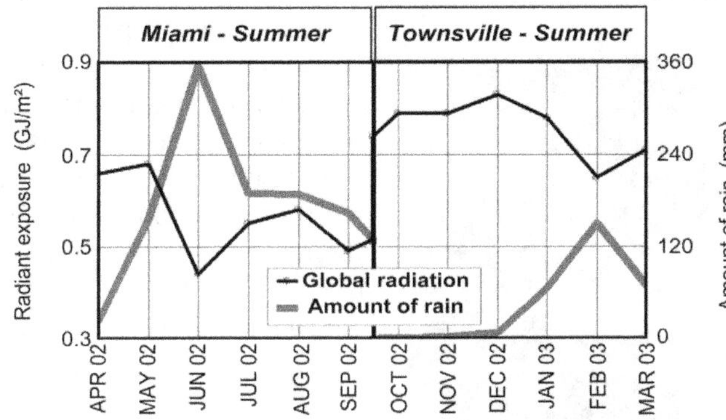

Figure 4.20: Monthly sums of global radiation and rain measured during the summer seasons in Miami/Florida and Townsville/Australia (adapted from [30, 50])

pares the two summer periods in terms of global radiation and relative humidity for the same 12-month period. In contrast to Townsville, the Miami summer reveals, besides lower global radiation, much higher moisture, as represented here by rainfall. Given the negative effects on the moisture conditions, it is not certain whether intensification of the radiation stress is reflected to the same extent in faster coating ageing, especially when humid-sensitive coatings are tested. It is unlikely that two years' conventional outdoor weathering in Miami, Florida, will be replaced by a one-year EverSummer Program. However, on account of the 1.5-fold increase in UV radiant exposure, a time-compressing effect can be expected and render the EverSummer Programme a sensitive method for intensifying outdoor weathering - unless the opposite is found in practice. At any rate, it should be tried out.

4.3.5 Outdoor weathering on the black box

Conventional outdoor weathering, as described in Chapter 4.3, involves working with test panels whose backs are open. With this kind of specimen mounting, a considerable part of the absorbed global radiation is dissipated to the ambient atmosphere via the backs of the test panels. The outflow of energy, primarily by convection, is promoted by free air circulating at the back. Limiting the air circulation at the back of the test panels can be expected to provide better simulation of normal use in specific cases.

The black box, a special specimen mounting, stemmed from the observation that the closed bodies of automobiles attain particularly high surface temperatures when parked in sunny places. In such cases, energy outflow by convection via the back is strongly limited due to a lack of free-air circulation in a completely closed car. The black box is predominantly designed to simulate the stress on automotive coatings. According to SAE J1976 [55], the black box is an empty, black-painted metal box (l x b x w: about 150 cm x 200 cm x 25 cm) mounted as a normal rack fixture, and commonly tilted towards the equator at an angle of 5°. The test panels themselves form the top area of the black box. Figure 4.21 shows the part of a weathering site equipped with a number of black boxes in southern Florida. Due to higher thermal stress, the black box should yield a higher rate of coating ageing than that provided by conventional outdoor exposure. The black box generates higher test panel temperatures during the daytime. It has been observed that the surface temperature of black-painted automotive coatings is up to 15 °C higher.

Although the black box is better at simulating the weathering stress in normal use, at least on automotive coatings, black-box weathering need not inevitably lead to higher time-compression than conventional outdoor weathering. As described in the discussions on the coating surface micro climate in Chapter 3.2.5, a higher surface temperature leads inevitably to less moisture on the test panel surface during outdoor exposure. A higher test panel surface temperature on the black box would be expected to lead to faster coating ageing. In contrast, the lower relative humidity on the surface of the test panels exposed on the black box would be expected to lower the ageing rate, especially during the time of highest irradiance. Although the black box was observed to prolong the nightly TOW, the question as to which of the two weathering stress factors prevails over the other cannot be answered in the absence of experimentation. The correct answer may depend on the type of coating material.

Figure 4.21: Test panels weathered on black boxes

The following two typical examples from the automotive coating test practice are to intended to answer the question: “What intensifying effect can be expected with the black box relative to conventional outdoor weathering". In the first example, specimens of two differently stabilised basecoat/clearcoat systems for automotive OEM were simultaneously exposed outdoors in Miami, Florida using conventional outdoor weathering (open back) and black-box weathering. The loss of gloss over the course of the three-year study is presented in Figure 4.22.

In this example, weathering on the black box led to a much higher loss of gloss after 18 months of exposure on the unstabilised clearcoat (#A-0) only (right diagram). The type of backing was not found to have any influence on the stabilised clearcoat (#A-0) for up to 36 months of exposure (left diagram).

The second example concerns studies carried out in Jacksonville, Florida. The idea was also to assess the influence of the backing on the weathering results. The studies were performed on identical series of 20 different basecoat/clearcoat systems. Two series were simultaneously weathered during the summer season of 2001, one conventionally, the other one on the black box. The results are shown in Figure 4.23. The degree of damage, visually assessed after exposure of 14 weeks, is plotted against the respective coating numbers (#01 to #20). The coating systems were arranged in increasing order of the damage caused by conventional outdoor weathering.

In this example, it was not the black box but rather the conventional outdoor weathering (open back) which led to higher coating damage on average. This result and the quite good ranking

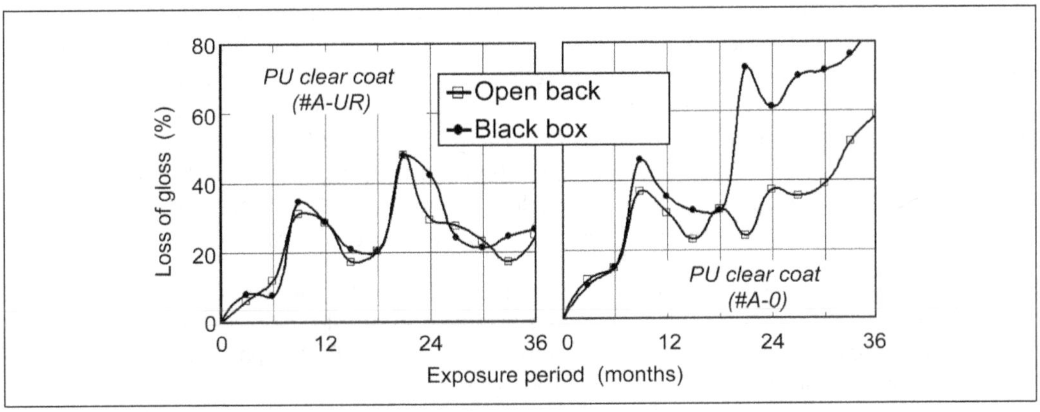

Figure 4.22: Loss of gloss by two differently stabilised clearcoats (#A-UR: stabilised, #A-0: unstabilised) on white-coloured basecoat, performed outdoors in Miami, Florida, as a function of exposure period and type of backing (adapted from [22])

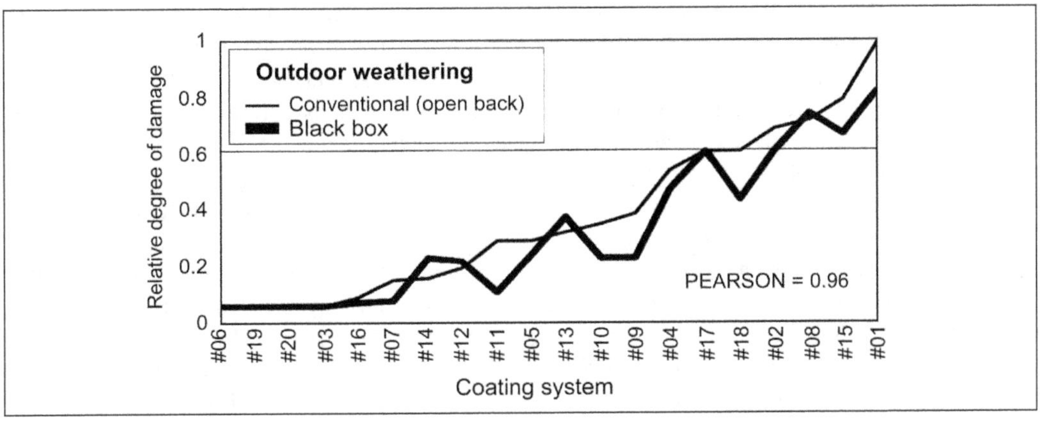

Figure 4.23: Visually assessed degree of damage on identical series, including 20 different automotive coatings, weathered outdoors in Jacksonville, Florida, for 14 weeks during the summer of 2001, as a function of the type of backing (from [45])

correlation (PEARSON = 0.96) between the two weathering tests means it makes no sense to replace conventional outdoor weathering by exposure on the black box - at any rate, not under the environmental conditions of Jacksonville, Florida, and in regard to automotive coatings. The two examples show that the decision as to whether **the black box is a sensitive alternative or a serious competitor to conventional outdoor weathering can be made only after appropriate preliminary studies.**

4.3.6 Outdoor weathering with additional wetting of the test panels

As has been emphasized several times, the poor repeatability of natural weathering stress is responsible for the poor repeatability of the results of outdoor weathering. Proposals have therefore been made to provide nature with a little assistance when designing the coating surface micro climate. Since global UV interacts synergistically with moisture, it would seem logical to retain the sun as radiation source, but to control the essential data of the coating surface micro climate, such as moisture and temperature.

Operators of weathering sites in Phoenix, Arizona, offer special weathering racks called spray racks for this purpose. A spray rack is equipped with nozzles to spray the test panel surface with water during weathering, continuously or cyclically. The frequency and duration of spraying are responsible for the water-related intensification of outdoor weathering. The film of flowing water on the test panel surface can be kept thin enough not to appreciably prevent global UV from reaching the coating. Therefore, the duration of spraying need not be temporally limited. This is particularly important for coatings whose worst case weather scenario is characterised by the action of global radiation on a wet coating surface. A weather situation frequently observed in Miami, Florida, during the summer in which the coating surface is covered by more or less extensive water puddles for as much as the whole day (see Figure 3.48) can be readily simulated by this type of intensified outdoor weathering carried out in Phoenix, Arizona.

Besides the artificially controlled moisture stress, the annual irradiance and average ambient temperature can also be seen as quite readily repeatable in dry desert areas, such as Phoenix, Arizona. The ready repeatability of the weathering stress is the first important advantage. This allows highly repeatable weathering results to be expected. Frequently, coatings that are moisture sensitive age the fastest in the aforementioned worst case weather scenario. Therefore, a considerable time-compressing effect, relative to conventional outdoor weathering, can be seen as another important advantage.

Since the duration of mutual action of global UV and water (TOW+U) is known to be the most effective period, permanent wetting of the coating surface should be done during the hours of maximum irradiance. Of course, the temperature of the water should not fall below ambient. As observed with certain polymeric materials, cyclic changes in phases with and without spraying may accelerate polymer ageing. There is no reliable evidence that such cyclical stress accelerates the photochemical ageing of coatings. As discussed in Chapter 3.2.3, damage by wetting and drying cycles, leading to alternating swelling and shrinking, were commonly seen as a result of loss of mechanical strength due to photochemical coating ageing. This was demonstrated by [38] on solid products made from PMMA. Additional cyclical spraying could make the state of coating ageing visible earlier. Nocturnal hours are a logical time to perform cyclical water stress. However, the most effective hours of maximum irradiance should not be interrupted by dry phases. Standardised methods generally recommend cyclical stress with spray phases which are too short or dry phases which are too long. Using these recommendations, the intensifying effect of mutually acting global UV and water can be expected to decrease to insignificance. In other words, spray phases presently recommended in the literature (for instance: 30 seconds every two hours during daylight hours [30]) can only be regarded as cleaning methods, but not as a means of accelerating coating ageing. The oft-recommended longer spray phases at night cannot be regarded as meaningful. A look at Figure 3.41 shows that clearcoats of common film thickness are saturated by water after about one hour. Only in the early morning hours could the water which penetrated into the coating at night contribute slightly to photochemical ageing.

Outdoor weathering with continuous wetting of the test panel surface during the hours of sunshine, as carried out in Phoenix, Arizona, can be recommended as a proper alternative to conventional outdoor weathering in Miami, Florida, because a greater time-compressing effect and better repeatability of the weathering results can be expected.

4.3.7 Outdoor weathering with controlled acid stress

The Southern Florida Acid Rain Test [56] is good example of how to improve the repeatability of outdoor weathering. The aim of developing the test was to suppress the spontaneity of acid atmospheric precipitation, as observed in Jacksonville, Florida. This property of acid atmospheric precipitation is generally held to be a possible reason for the poor repeatability of Jacksonville outdoor weathering. Therefore, it was proposed that outdoor weathering be carried out not in Jacksonville, Florida, but at another weathering site having roughly the same climate, but without any stress due to acid atmospheric precipitation. The missing natural acid stress is replaced by artificial acid rain sprayed in controlled fashion onto the surface of the test panels. A mixture of sulphuric, nitric, and hydrochloric acids, with a mass ratio of 1 : 0.3 : 0.17, diluted with water to pH 2.5, serves as artificial acid rain. During exposure, the test panels are sprayed with the acid mixture two times a week.

The test panels are mounted on conventional weathering racks and exposed outdoors for 14 weeks during summer (June to Sept.) in the same way as conventional outdoor weathering is performed in Jacksonville, Florida. As already described in Chapter 3.2.6, a synergistic interaction of acid with the other weathering stresses proved to be responsible for the typical acid etching on the coating surfaces. As experienced at the Jacksonville outdoor weathering, only the climate of the summer season provided the necessary stress to trigger the aforementioned synergy. **Therefore, the Southern Florida Acid Rain Test can be successfully performed only once a year, from June to September!**

The following example compares the results of outdoor weathering carried out under natural and artificial acid stress. The studies were performed on identical series of 20 different basecoat/clearcoat systems for automotive OEM. Two series were simultaneously exposed outdoors during the summer season of 2001, one series in Jacksonville, Florida, while the other was exposed Southern

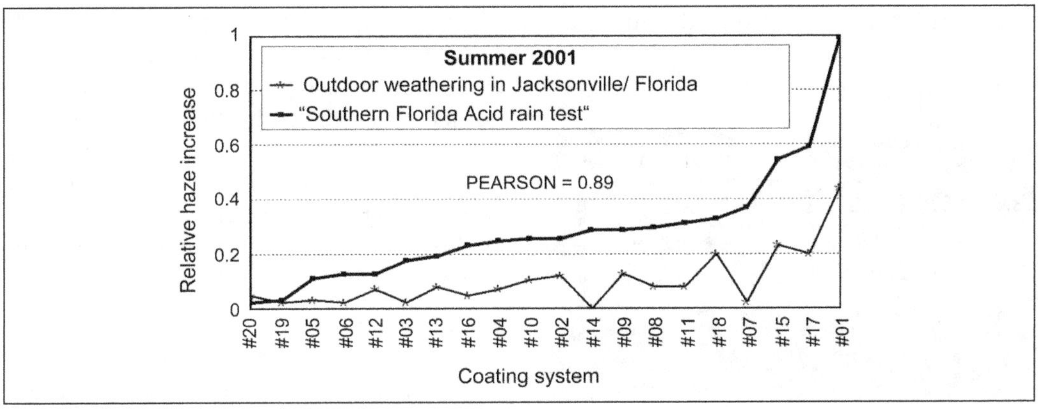

Figure 4.24: Haze increase caused by conventional outdoor weathering in Jacksonville, Florida, compared to that caused by the Southern Florida Acid Rain Test performed in Homestead, South Florida (from [56])

Florida Acid Rain Test in Homestead, Florida. The results are shown in Figure 4.24. The haze increase measured after 14 weeks' exposure is plotted against the respective coating numbers (#01 to #20). The coating systems were arranged in increasing order of the damage that arose in Southern Florida Acid Rain Test. Haze increase proved to be a good objective measure for assessing acid-related coating damage. Haze increase correlates well with the visually assessed damage, and, therefore, serves as a suitable ageing criterion. Comparison of the two tests shows that their results differ in the degree of damage, but that they correlate quite well with each other, as expressed by a PEARSON of 0.89. **Due to this good ranking correlation, the Southern Florida Acid Rain Test is a proper alternative to the conventional outdoor weathering in Jacksonville, Florida, as a test of the weatherability of coatings intended for use in special regions in which stress by acid atmospheric precipitation is part of normal use.**

Furthermore, the use of a weathering site not stressed by acid atmospheric precipitation reveals a double advantage. Testing the weatherability both with and without acid stress can be performed at the same weathering site. Unfortunately, the influence of the other climatological data, such as temperature, moisture and irradiance, on the weathering results could not be suppressed by controlling only one of the stress factors. Repeating the Southern Florida Acid Rain Test in the following years on the same coating series at the same weathering site and during the same season revealed a fluctuation in results that was similar to that found in Jacksonville, Florida. This produced the additional experience that the poor repeatability of the outdoor weathering in Jacksonville, Florida, is caused not only by the spontaneity of acid atmospheric precipitation, but also by the yearly changes in water, heat and irradiance stress.

4.3.8 Irradiation and weathering behind glass

As described in Chapter 4.3.6, the spray rack allows the TOW and TOW+U during outdoor weathering to be controlled. To control further climatological parameters of the coating surface micro climate, such as temperature and relative humidity, it is necessary to completely isolate the test panels from the ambient climate. If the sun is to be kept as radiation source, isolation must be effected by glass which is transparent to the total range of global radiation (300 to 2500 nm). Operators of weathering sites offer racks equipped with special apparatus for performing irradiation or weathering under conditions of controlled temperature and relative humidity. Only the irradiance is left as uncontrolled parameter. Figure 4.25 shows typical equipment used at a weathering site in Phoenix, Arizona.

Where such equipment is to be used for testing interior automotive components, corresponding windscreen glass is used for isolating the test panels from the ambient climate. In the event that the coating surface micro climate does not have to be controlled, the equipment is typically used for irradiation, as described in Chapter 2.1. The coating surface micro climate regulates itself as a function of the irradiance of the incident global radiation. The surface of black-painted test panels irradiated in this way in Phoenix, Arizona, is reported to reach temperatures up to 120 °C. Since coated materials are seldom used as interior automotive components, this kind of irradiation has acquired little significance in the testing of coatings.

Figure 4.25: Equipment for weathering outdoors behind glass (adapted from [30])

Equipment with windows which are highly transparent to total global radiation can be seen as a proper means of further improving the repeatability of outdoor weathering, because only the irradiance remains uncontrolled. When

Phoenix, Arizona, is the chosen weathering site, the fluctuation in annual radiant exposure, as shown in Figure 4.3, is of minor significance and so has only a weak influence on repeatability. **Total isolation from the ambient climate, combined with the option of humidifying and controlling the temperature of the climate behind the window, would allow the test to be performed on all climate variants that have proven capable of intensifying outdoor weathering.** Special UV-transparent window materials are commercially available (for instance: quartz, special glass, and PVDF film).

Figure 4.26: Equipment for weathering outdoors behind glass in a controlled climate and air pollutant stress

The equipment described in [57] can be seen as another typical example of the use of outdoor weathering behind glass in a controlled climate and the sun as radiation source. It serves to test the weatherability of polymeric materials, including coatings, where normal use includes attack by aggressive air pollutants, such as SO_2, NO_x, O_3, and Cl_2. Figure 4.26 shows the corresponding weathering equipment.

A gas-metering device subjects the test panels to an airflow having a controlled content of pollutants, controlled relative humidity and controlled temperature. The air flows permanently along the slit between the window and the test panel surface. However, this method of outdoor weathering has not achieved any significance for testing coatings and therefore is mostly used for research.

4.4 Reference coating for outdoor weathering

The repeatability of weathering stress and the weathering results are generally poor, as has been mentioned several times in this book. Therefore, it has become common practice to use coatings as reference materials whose weatherability is known and was established in previous outdoor weathering tests. It is also standard practice for the test engineer to choose the appropriate coatings, mostly from his own series production. Panels of the chosen reference coatings are weathered simultaneously with the coatings to be tested. Presently, this method is the best way of assessing the repeatability of the weathering stress. Reference coatings can also detect possible synergies between individual weathering stresses.

Tracking the weathering-induced damage of the reference panels from test to test enables the repeatability to be more reliably assessed than is possible by simply comparing the respective climatological data. An advantage of reference coatings chosen in this way is the close proximity to the coatings to be tested. However, reference coatings have to be exchanged if new types of coatings have to be tested. Furthermore, reference coatings have to fulfil all the requirements imposed on a reference material in general. Appropriate reference panels have to retain their properties over long periods in storage. They must be produced and stored in sufficient quantity. The ageing behaviour of reference coatings must not alter under corresponding conditions of storage. Therefore, they are often stored in the dark at low temperatures. Reference coatings have to reproduce all their properties when they are used to produce new panels.

As is sometimes observed during repeats of outdoor weathering tests, differences in the coating's ageing behaviour occur from test to test, considerably contradicting the differences in the respective climatological data. Furthermore, some weathering sites do not provide the possibility of measuring the essential climatological data during the test. So, it is often not possible to assess the repeatability by means of such data and the use of reference coatings remains the only way of assessing the repeatability. For Jacksonville, Florida, for instance, the available climatological data are measured at Jacksonville International Airport (JAX) [58], which is about 40 km away from the weathering site. Therefore, outdoor weathering in Jacksonville, Florida, performed without corresponding reference panels cannot be deemed sensitive. Exposing an appropriate number of reference panels simultaneously with and among the panels to be tested has proved to be indispensable for testing automotive coatings in Jacksonville, Florida. In this case, at least three coatings of known sensitivity to acid attack, previously proven in Jacksonville, Florida, should serve as reference panels.

In view of the shear diversity of coatings for different products and purposes, it is not hard to see that operators of weathering sites are unable to provide the appropriate reference coatings. This raises the question: "What about the reference materials commonly employed for weathering polymeric materials, e.g. blue wool?" This material is a standardised blue coloured fabric and therefore unsuitable for outdoors. Another reference material is the polystyrene light-fastness standard, delivered in the form of solid plastic chips. Polystyrene chips are often used for outdoor weathering because they are easy to handle. Special film made from polysulphone or polyethylene are available for this purpose, too. These reference materials are fundamentally different from coatings and therefore have not gained appreciable significance in coatings testing.

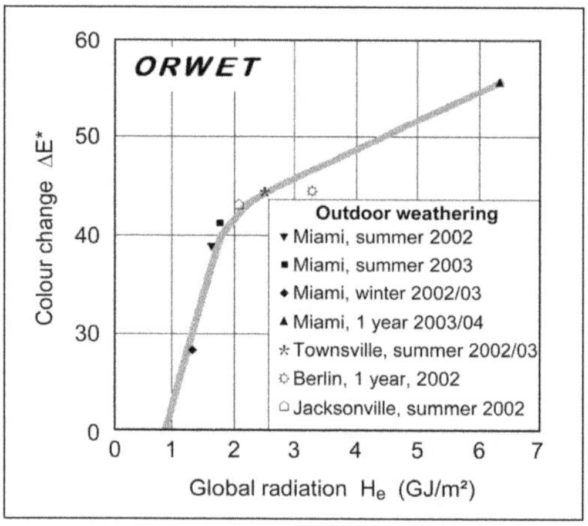

Figure 4.27: Colour change in the reference coating ORWET caused by conventional outdoor weathering at different weathering sites, during different seasons, with different exposure periods, as a function of the radiant exposure

The only reference material that can serve as a generally useful reference coating suitable for outdoor weathering is the "Orange Weathering Tile" (ORWET). ORWET is commercially available from EMPA Switzerland [59]. It is a simple melamine-based coating that contains an orange-coloured, highly radiation-sensitive organic pigment (CIBA-"Irgalith" Orange F2G). ORWET is usually painted onto 0.4 mm-thick aluminium panels in layers about 100 µm thick.

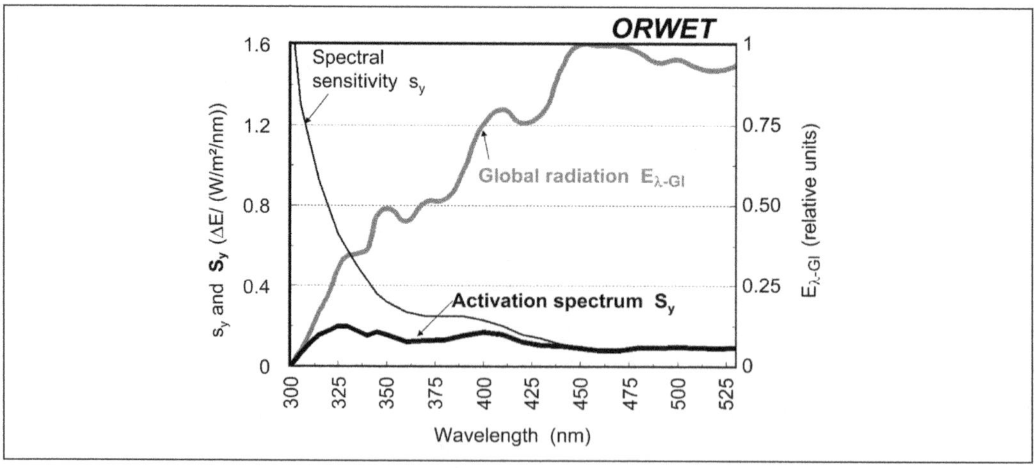

Figure 4.28: Activation spectrum of global radiation (S_y) acting on reference coating ORWET, resulting from the spectral sensitivity of the ORWET (s_y) and the spectral distribution of the global radiation ($E_{\lambda\text{-GL}}$), according to CIE 85, Table 4, calculated for the wavelength from 300 to 500 nm.

Figure 4.27 shows the change in colour as the corresponding ageing criterion from strongly orange up to almost white, as a function of the radiant exposure (total global radiation) after outdoor weathering at different sites during different seasons and for different exposure periods. As can be seen, the joint curve is almost directly proportional to the radiant exposure up to about 2 GJ/m^2. Above this value, the curve becomes a regressive exponential curve. As can be seen, all the colour change values agree well with the joint curve in spite of the high diversity of the outdoor weathering tests. In other words, ORWET appears to be hardly influenced by the other weathering stresses, such as heat and water. Acid atmospheric precipitation contributes only weakly to coating ageing. ORWET has therefore proved to be a suitable reference coating for detecting small differences in the global radiation stress. Unfortunately, ORWET is not a convenient vehicle for describing all the weathering stress occurring during outdoor weathering. The relatively good repeatability of the colour change, recognizable from the two tests carried out in Miami, Florida, during the summers of 2002 and 2003, is probably also due to its sensitivity to global radiation. Figure 4.27 can thus also be used to interpolate the radiant exposure by means of ORWET's weathering-induced colour change. This is important for weathering sites which have not the technology for measuring radiant exposure. Further studies have revealed the reason for ORWET's high sensitivity to global radiation.

Figure 4.28 shows the curves for the spectral sensitivity of ORWET (s_y), the spectral distribution of the global radiation ($E_{\lambda\text{-Gl}}$) according to CIE 85, and the activation spectrum (S_y) resulting from the two aforementioned curves, as described in Chapter 3.2.1.2. The wavelength range from 300 to 530 nm was chosen for this presentation. It is a typical feature of coatings containing organic pigments that the activation spectrum ranges into the VIS and shows several clear maxima. In contrast, common activation spectra for polymers, including the common paint resins (see Figure 3.20), show only one maximum, near 320 nm, and no sensitivity to radiation above 400 nm. In the case of ORWET, the VIS range of the global radiation contributes considerably to coating ageing. VIS is responsible for about 50% of the colour change.

Now, let us address the question: "How to use ORWET for outdoor weathering?" A small piece of ORWET measuring 1.5 x 1.5 cm (depending on the available colour measuring technique) is glued on a corner of the test panel, as shown in Figure 4.29. This is sufficient to detect the stress by global radiation during outdoor weathering. It is very important for the piece of ORWET not to be covered by elements for affixing the test panels or by other things.

Especially for weathering sites that have no way of permanently supervising the exposed test panels, OEWET can provide the evidence as to whether all the test panels were exposed to the sun in the correct way. There have been occasions when certain coatings exhibited unusually good weatherability that could not be explained when they were returned from the outdoor weathering. Clearly, such test panels had little or no exposure to the sun during the weathering period.

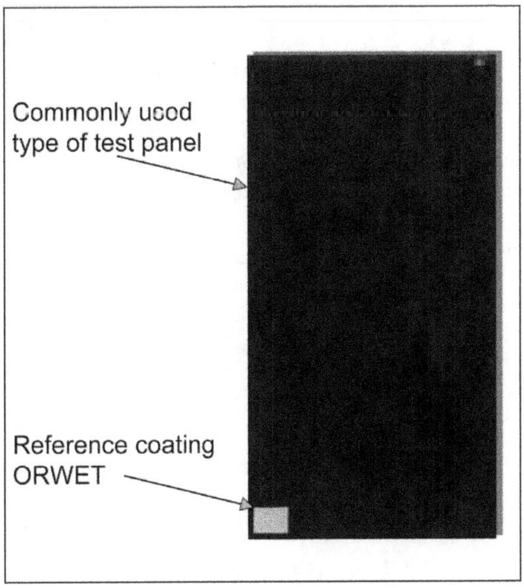

Figure 4.29: Photograph of a black-painted test panel (150 mm x 300 mm, as commonly used for outdoor weathering of automotive coatings) with a small piece of OEWET (1.5 mm x 1.5 mm) glued on the bottom left hand corner

4.5 Starting date and duration of outdoor exposure

It can be stated in general that the weather is incapable of repeating itself and each outdoor weathering test is unique. Due to the poor repeatability of outdoor weathering, the weathering results must always be expected to vary from test to test. Statements about weatherability or service life predictions that are based only on one test or a test that was too short are always accompanied by a high degree of uncertainty. For this reason, the outdoor weathering should always be repeated at least once. It is common for outdoor weathering tests to be performed without interruption for several years. However, as has often been observed, the result of a two-year test need not be identical with the averaged results of two tests, performed separately in one year, starting with non-weathered specimens each time. The rate of ageing often changes in the course of weathering. If little time is available for testing, one-year outdoor weathering tests should be performed simultaneously at several weathering sites of identical or similar climate. Proposals to this end are provided in Tables 4.1, 4.2, and 4.3. In such cases, a higher number of tests will improve the confidence of predictions made based on the results.

The following example shows how the exposure period (duration of weathering) can considerably influence the coatings' weatherability ranking, in this case within a series of 10 different aircraft coatings. Figure 4.30 shows the loss of gloss by the coatings over six-year outdoor weathering in a conurbation stressed additionally by aggressive air pollutants, in Leipzig, Germany.

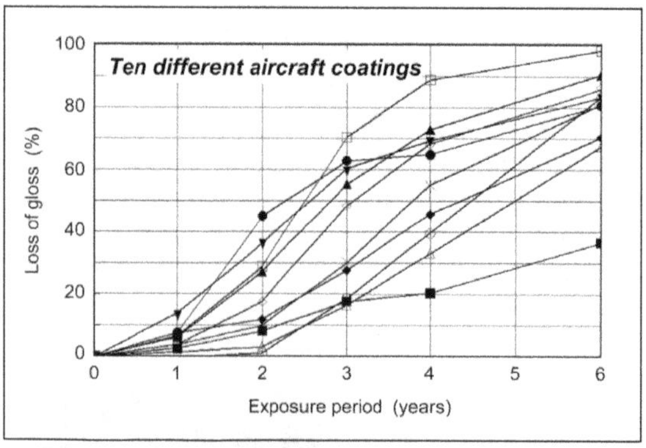

Figure 4.30: Loss of gloss (20) by ten different aircraft coatings during a 6-year conventional outdoor weathering in an urban area in Leipzig, Germany (from [60])

Due to weathering-induced increase in surface roughness, the loss of gloss increases steadily. The curves for the 10 coatings are indicated by differently shaped points. As can be seen in Figure 4.30, the curves cross each other several times during outdoor weathering. This proves that the weatherability ranking inside the series changes considerably from year to year during outdoor weathering. A virtually permanent ranking was reached only after 4 years. In other words, in this case, four years of outdoor weathering proved necessary before sufficient information for choosing the best aircraft coating was acquired.

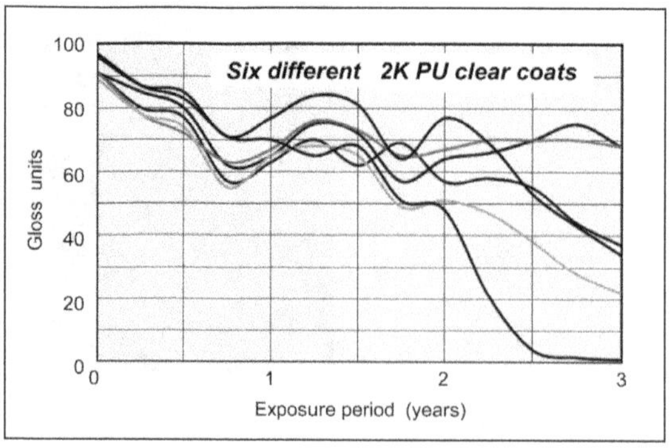

Figure 4.31: Gloss (20°) of six white-coloured automotive coatings with differently treated 2-pack polyurethane clearcoats during three-year conventional outdoor weathering in Miami, Florida, that commenced in January 2002 (adapted from [22])

The following example yielded the same conclusions. It proves that today's automotive coatings also need several years of outdoor weathering before newly developed or changed coating systems can be reliably assessed.

Figure 4.31 shows the change in gloss of six different basecoat/clearcoat systems for automotive OEM during three-year conventional outdoor weathering in Miami, Florida (5° south). The wavy shape of the gloss curves is typical of modern clearcoats. The steep downward slope of the curve characterises the incipient failure of the respective coating. The exposure period to 50 % loss of gloss is a standard criterion for describing the coating's service life. In this example, the curves of the tested six coatings revealed clear differences in weatherability only after 30 months of exposure.

The wavy shape of the gloss curves raises the question "On which date of the year should outdoor weathering tests be started?" There is only one correct answer: "Outdoor weathering must always start as close as possible to the same date of the year". This holds for all kinds of coatings to be tested outdoors, especially when only periods of less than one year are necessary or available. The only acceptable exception is when the influence of different scheduled starts is to be studied. **However, it can generally be expected that a failure curve produced by outdoor exposure initiated in June will not be the same as a failure curve derived from outdoor exposure initiated in December.**

In special cases, outdoor weathering only makes sense when performed during the summer, such as in Jacksonville, Florida. Here, it is imperative that the test always be commenced in the same week of the year. Although, in this case, only 14 weeks are usually needed for outdoor weathering, nevertheless, only one test a year can be performed. As has been discussed in Chapter 4.2.3, at least two repetitions are recommended for obtaining sufficient information to permit reliable assessment of the coatings' weatherability in regions where acid atmospheric precipitation is part of weathering stress. The average of three tests usually forms the basis for assessing weatherability. In such cases, those coatings which have proved to be the best, based on the average damage, must never have exceeded an agreed acceptable degree of damage in any of the three tests performed separately.

4.6 Recommendations for performing outdoor weathering

This Chapter of Chapter 4 is intended to provide users of outdoor weathering services with recommendations on the preparation, performing, and post-processing of ordered weathering tests by dealing with frequently asked questions. The question often arises: **"Which shape of specimen is the best with regard to intended 'normal use'?"** Sometimes, completely furnished end-products coated with the coating to be tested can be observed on weathering sites. However, for testing coatings in development or in current series production, less complicated specimens are sufficient. Plain panels are the optimum shape of specimens for coatings testing. They can easily be mounted on common weathering racks. Implementing and retaining exact exposure angles over the whole specimen surface do not present any problems. With regard to the assessment of weathering-related coating ageing, only a plain test panel permits correct measurements of optical surface properties, such as gloss, haze, and colour. Visual assessment of coating ageing is also supported by the plain shape of such test panels. The entire test method, starting with the production of the test panels, followed by their preparation, pre-treatment, transport, outdoor exposure, after-treatment, measurement, assessment, up to archiving, is favoured by plain test panels.

What about the size of the test panels? Certain heterogeneities caused by application, transport, handling and unexpected influences during weathering (for instance, bird excrement) cannot be ruled out. Therefore, test panels should always be big enough for correct assessments. According to DIN 1514 [61], the weathered surface of a test panel should not be smaller than $0.03\ m^2$. The length of one of the edges should not be shorter than 100 mm. Since the coating damage is often determined visually, test panels measuring 30 cm x 15 cm are recommended for outdoor weathering of automotive coatings. Disturbing influences can arise from the edges if overly small test panels are used. Proper edge covering will be impossible if mild steel serves as substrate.

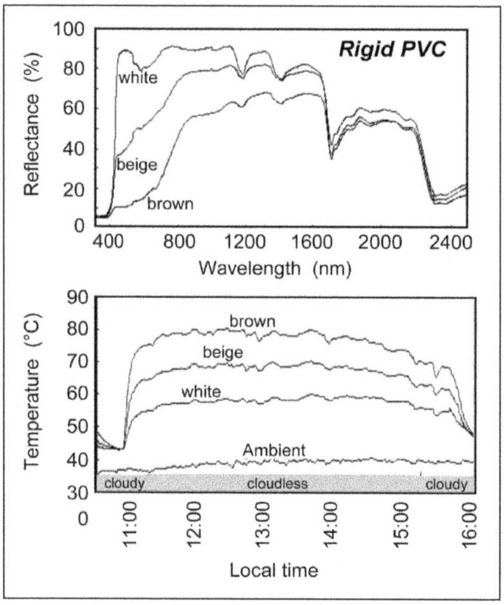

Figure 4.32: Relationship between colour, reflectance, cloudiness and surface temperature of three differently coloured PVC specimens during conventional outdoor weathering on a typical June day, in Phoenix, Arizona (from [29])

Steel must be protected against rusting unless stainless steel is used. However, bear in mind that edge covering can reveal incompatibility with the coating material! Heterogeneities in coating ageing can also be caused by differences in the coating surface micro climate over the whole weathering rack if the test panels are overly small. As observed on test panels measuring 5 cm x 10 cm and made of black-painted aluminium sheet material, and conventionally exposed on a common, completely filled weathering rack outdoors in Miami, Florida, the coating surface temperature varies by 5 °C according to whether the test panels were mounted on the centre or the edge of the weathering rack.

Another important question is **"Which coating colour should be used for outdoor weathering?"** As has been discussed in the context of the coating surface micro climate in Chapter 3.2.5, different colours lead to different surface temperatures. This can yield different ageing rates. The lower diagram of Figure 4.32 shows how strongly the surface temperature of specimens made from rigid PVC and exposed outdoors in Phoenix, Arizona, differs as a function of the colour and the local time of day. The upper diagram shows the reflectance of the surface of the same specimens against the total range of global radiation. The brown-coloured specimen reached the highest temperature (up to 80 °C) due to its lowest reflectance, especially in the VIS and NIR. Furthermore, the diagram shows that, during cloudy hours, in this case before 11 a.m., the colour had no discernible influence. With regard to global UV, the colour did not influence the reflectance.

If the weathering focuses on **testing colour fastness**, which is a concern for all pigmented or otherwise coloured coatings, such as topcoats, façade paints, basecoats, and others, it is not possible of course to use uniformly coloured samples for the purpose of testing. Organic pigments, especially, are vulnerable to weathering - just like coating resins. With basecoat/clearcoat paint systems, weathering testing predominantly focuses on the weathering behaviour of the clearcoat. As it constitutes the outermost coating layer, the clearcoat is directly exposed to all weathering stresses.

In order to avoid differences in the coating surface micro climate during outdoor weathering, uniformly coloured basecoats are recommended. Two types of clearcoats (e.g. a TSA and a polyurethane coat), applied to two differently coloured basecoats (e.g. black and white) have proved sufficient. This kind of selection is generally recommended, for instance, when the qualification, compatibility, and efficiency of clearcoat additives are being studied. For testing basecoat/clearcoat systems in Jacksonville, Florida, black test panels prove to be more suitable than white ones. Black test panels reveal surface damage by acid etching much more clearly than their white counterparts. They are often preferred to white ones when automotive coatings are being tested. However, what is true in Jacksonville, Florida, need not apply to all other weathering sites. As has been discussed in Chapter 3.2.5, black does not always represent the worst case concerning the test panels' colour.

Note that a higher surface temperature leads to lower moisture in the coating surface micro climate. White basecoats are recommended if, besides the clearcoat's ageing behaviour, their protective effect on the basecoat is to be studied. White specimens allow coating ageing to be tracked through changes in colour. However, the use of uniformly coloured basecoats does not inevitably mean that the basecoats have to be of uniform type as well. Such kind of uniformity can reveal incompatibility between basecoat and clearcoat during weathering that leads to ageing effects, such as delamination and colour anomalies that will never occur in normal use. When multi-coat systems are tested, the panels should be coated with the full coating system that will be employed in series production.

What about the substrates for the test panels? If possible, corrosion-resistant materials should be used. Substrates of aluminium have proved suitable, even in the aggressive climate of Jacksonville, Florida, for testing automotive and aircraft coatings. Aluminium substrates do not require any of the edge covering that is required when mild steel is used. The thickness of the test panels should ensure sure sufficient mechanical stability to avoid loss of evenness (at least 1 mm for aluminium). Coatings for façades, concrete, or plaster should be applied to the actual substrate. Test panels for these reasons should not be thinner than 5 cm in order that the real coating surface micro climate may be simulated correctly. Such substrates require that special attention be paid to edge covering and correct conditions on the back.

With wood varnishing, the substrate should be made of wood pre-treated by the methods and means normally encountered in normal use. Weatherability of scratch-proof coatings or other special layers intended to protect certain substrates against mechanical damage should be tested on the real substrates (for instance: PC for car window glass). It has been observed that even highly weather-resistant scratch-proof coatings have failed when weathered in conjunction with the right substrate. This can occur if the scratch-proof coating does not additionally protect the substrate against the global UV. For instance, in the case of covered PC, under-chalking caused by degradation of the PC substrate on the interfacial area could result in delamination.

Special studies of clearcoats require that the **weathering tests** be performed **on free films**. Certain methods of assessing coating ageing can be best done on free films. A special substrate proved suitable for outdoor weathering of free films of clearcoats for automotive coatings carried out in Jacksonville, Florida. Before weathering, the free films were immobilised in direct contact with the surface of common test panels, which were treated with the same clearcoat. In other words, the film and substrate consisted of the same clearcoat. Incompatibility between clearcoat film and substrate during weathering were ruled out by this kind of mounting. It is not recommended that the test panels be fitted with special heat-insulating sheets on the back with a view to enhancing the surface temperature during weathering, as the coating surface micro climate can be influenced in an unpredictable manner. A time-compressing effect relative to conventional outdoor weathering (open back) cannot be expected.

Another important question is: **"How long should you wait after producing the test panels?"** As has been described in Chapter 3.1.5, the coating may not have completely lost its solvent after application. In another case, the curing process may not be completely finished. As is sometimes observed, freshly coated test panels stick to each other if stored or transported such that the coating surface of one test panel is in direct contact with the coating surface of another. You should wait a certain amount of time after the test panels have been produced. Test panels of automotive coatings should be stored carefully for at least six weeks at normal room temperature or slightly higher (up to 50 °C). The coating surfaces must never be allowed to touch!

Furthermore, it cannot be ruled out that the ageing behaviour undergoes marked changes during storage, due to a gradual change in glass-transition temperature.

Figure 4.33: Test panels mounted on typical weathering racks that are adjustable to different exposure angles at a weathering site in South Florida

What about the weathering racks? Figure 4.33 shows a series of typical standardised weathering racks fully equipped with differently coloured test panels of the recommended size. The equipment for altering the exposure angle can be clearly seen

Besides sufficient mechanical stability of the rack, the test panels must be securely immobilised! This will be particularly important if the weathering site is not continually supervised by staff. The test panels might become detached from the rack and be lost, especially at weathering sites located in regions often at risk of storms. This may lead to complete failure of the outdoor weathering test and delay the development of the coatings concerned by one year, at least. In Miami, Florida, for instance, special emergency programmes for avoiding loss of test panels are triggered if a hurricane threatens. Ensure that weathering racks are made from material which are not sensitive to corrosion! This is necessary to prevent products of corrosion, such as rust, from reaching and influencing the test panels.

When different coatings have to be compared with each other or reference materials, simultaneous exposure of the concerned test panels on the same weathering rack is recommended. This avoids discrepancies caused by differences in air circulation or the type of ground beneath the rack, although there is no sure evidence that the type of ground has any major effect on the rate of ageing. Applicable standards recommend a minimum distance to the ground of 0.5 m. The ground should be grown with vegetation typical of the character of the respective weathering site. Easy adjustment of the desired exposure angles should be possible. Deviations of $\pm 2°$ should not be exceeded. Ensure that the diffuse portion of the global radiation is optimally used for weathering. For horizontal adjustment, the samples should face the sky at an angle ranging from 20 to 160° relative to the horizon. Take care that specially shaped test specimens exposed on the same rack do not cast a shadow on other test panels!

Furthermore, the user of a weathering service should ask the operator of the used weathering site about changes in special environmental conditions, such as the type of ground, stress by dust, traffic, air pollutants, vicinity of power stations or industrial plants, biological situation and the like since the preceding outdoor weathering tests.

"Take great care throughout the steps of preparation, pre-treatment, performance, and post-treatment" is another important recommendation. As mentioned before, each failed outdoor weathering test means the loss of one year at least in coating development, or greater uncertainty in the monitoring of any current series production. Outdoor weathering costs time and money. Care begins with weatherproofed labelling of the test panels, especially when series of uniformly coloured coatings are tested. The state of the test panels before weathering should be thoroughly recorded.

Certain surface defects, for instance, caused by prehistory, often cannot be totally avoided, especially when the test panels are produced in such laboratories which have no facilities for series production. Such defects can lead to damage which is atypical of the coating to be tested. That can lead to a wrong assessment about the weatherability of the coating. Areas on the test panel that deviate from normal appearance should be photographed. As is frequently observed, blisters, pitting, craters, or scratches, that were present before weathering are wrongly included in the assessment of the weathering result. Ensure that the measuring points, where magnified photographs were taken or optical properties were measured before weathering, are permanently marked in order that the studies after weathering may be done on the same measuring points. Note that certain parts of the test panels may have lost their evenness during weathering. Correct measuring of the optical properties on such measuring spots is not possible, even where only small deviations in evenness have arisen. Therefore, a corresponding number of measuring points on the test panel surfaces should be prepared in order that the basic requirements of statistics may be met. Additionally, attack by bird excrement, hailstones, or scratching can lead to loss of useable measuring points.

What about transporting the test panels to and from the weathering site? Note that the scratch resistance may be considerably reduced by weathering! The weathering-induced reduction in scratch resistance should not be altered by the transport conditions on the way from the weathering site to the laboratory performing the measurements. Carelessness during transport of the test panels to and from the weathering site has often led to serious difficulties in assessing the weathering results due to strongly scratched parts of the test panel surface. Only special transport containers that reliably prevent test panels from touching each other have proved suitable. Furthermore, any contact with wrapping material should be avoided. A distance of several millimetres between one test panel and another is essential. In order to obtain the most information about the outdoor weathering test, such as attack by dust, salts, tar spots, bird excrement, leaves, algae, fungi, it is recommended that the test panels be returned to the client's laboratory without any kind of after-treatment. That means without washing, brushing or polishing, i.e. in the state in which they were taken from the weathering rack. Note that the dirty test panels still need special care to avoid being scratch by attached sand and other substances! During all post-treatment steps, the test engineer should keep an eye on his test panels for as long as possible, at least during cleaning.

"What about cleaning the test panels after outdoor weathering?" Carelessness during cleaning may lead to additional damage that is mostly indistinguishable from the weathering-induced damage. Good cleaning is necessary for correctly determining the weathering-induced damage visually, as well as by optical measurement. The following cleaning method has proved suitable for test panels of automotive coatings after outdoor weathering.

Cleaning steps

The first step is to store the test panels in 40 °C in demineralised water for 2 days. The purpose of this step is to dissolve salt crusts, e.g. crusts of calcium sulphate on coatings exposed in Jacksonville, Florida. Water-soluble salt crusts can easily be confused with acid etching. For test panels exposed in Miami, Florida, the first step is to facilitate the removal of mycelium fungi, which is sometimes observed after exposure on other weathering sites with humid-hot climate.

The second step is to wash the test panels. Washing should start off gently under normal tap water and finish in a harsh manner with a sponge and surfactant-containing water.

The third step is to polish the test panel surface with an organic, non-abrasive polish, and a microfibre tissue. Its purpose is to remove tar flecks (often observed after exposure in the harbour area of Jacksonville, Florida) or obstinate residues of fungi mycelium. It is necessary to clearly identify damage caused by algae and fungi (pinholes).

All cleaning steps should be documented with photographs. Before and after each cleaning step, photographs should be always taken of the same measuring points on the test panel surface. Operators of weathering sites frequently offer the users of the weathering service the chance to wash and assess their test panels at regular periods during outdoor weathering as a way of tracking coating ageing. The main reason is to record the date of the coating's breakdown if it is expected before the scheduled exposure period ends. However, on account of the aforementioned problems, disruptions to the course of the outdoor weathering test should be avoided, particularly if cleaning processes are involved. If such interim results are desired nonetheless, the outdoor weathering test should be started with a higher number of test panels for each of the coatings to be tested. During the scheduled course of the test, panels could be removed after different exposure periods. Another method that is sometimes employed consists in covering part of the test panel surface by a material which is resistant to weather, is waterproof, and is not transparent to light. In this case, only the covered part is thermally stressed. This may help to distinguish between thermally-induced and radiation-induced ageing effects.

4.7 Advantages and disadvantages of outdoor weathering

The advantages and disadvantages of outdoor weathering are juxtaposed in Table 4.9.

One of the essential advantages is the closeness of the test to reality. Natural weathering stress is used for testing. According to the criteria for accelerated testing mentioned in Chapter 2.5, a time-compressing effect is gained by a higher frequency of the most effective exposure conditions. The natural extremes of the weathering stresses are not exceeded. The coatings to be tested are subjected only to stresses which can also occur in "normal use". All ageing processes that occur in normal use also occur during outdoor weathering, but at a higher rate. Certain ageing processes will only occur if the weathering stresses combine in the same manner as found outdoors. On the other hand, the use of the natural environment protects against ageing processes that never occur in practice. **Due to its closeness to reality, outdoor weathering will never lose its justification.**

Table 4.9: Advantages and disadvantages of outdoor weathering

Advantages	
Close vicinity to the reality (normal use)	Natural exposure conditions, no unrealistically exceeded weathering stresses, practice-relevant coating ageing
Economically best weathering method	Low running costs (no costs for irradiation), low technical expenses/low costs for investment, low priced weathering service is possible
Disadvantages	
Insufficient repeatability and reproducibility of the weathering results	Considerable variation of the climatological data (weathering stresses) from year to year, local-related differences of the climatological data, even in regions with same climate
Poor time-saving effect (highly time-consuming method)	Irrational utilization of the time disposable for weathering tests: ineffective night hours, poor-effective morning and evening hours, poor-effective cool and cold seasons, loss of radiation by clouds
Possible loss of time because of enhanced probability of outfalls	Loss or severe damage of test panels caused by storms, hurricanes, or not allowed interferences if the weathering site is not permanently supervised
Considerable uncertainty of the test results	Possible misinterpretation of weathering results caused by unexpected or unrecognised environmental influences being untypical for the used weathering site. (e.g. caused by building works, traffic, averages, biological attacks)
No possibility of permanent specimen observation and exercising influence on the weathering course by the users of weathering service	No information about accruement of damage, no possibility to make a decision immediately (e.g. stopping the test prematurely or not)
No immediate restart after outfall or interruption of the weathering exposure	No temporal constancy of the climatological data in the course of the year (different results between interrupted and non-interrupted weathering tests can be expected in spite of identical exposure period)

Outdoor weathering will always be the last hurdle for a newly developed coating prior to series production. Outdoor weathering offers some economic advantages. Using the natural environment does not incur costs for irradiation, heat, humidity, and rain. The very low energy costs are a major advantage, especially relative to artificial weathering by means of xenon lamps, which consume high amounts of energy. Furthermore, the weathering racks for the most commonly employed conventional outdoor weathering cannot be seen as cost-intensive investment. Therefore, outdoor weathering services can be offered at relatively low cost. **Calculated per square metre of test panel surface, outdoor weathering will always substantially undercut the costs of all other methods of weathering.**

The disadvantages of outdoor weathering have been discussed in Chapter 3.2.1.4 und 4.5 in more detail. Poor repeatability and reproducibility of weathering stress and therefore of the weathering results are generally seen as the major weakness. The weather is incapable of repeating itself from year to year (see Figures 3.27 to 3.29 and 4.03) and reproducing itself from one location to the other (see Tables 4.1 to 4.5). As commonly used in the context of outdoor weathering, repeatability refers to weathering tests repeated at the same weathering site, while reproducibility refers to weathering tests repeated at different weathering sites.

The relatively little time-compressing effect relative to normal use can be seen as a further essential disadvantage of this kind of accelerated testing. Outdoor weathering is sometimes called long-term weathering in contrast to the discussions in Chapter 4.1. Low time-compression means high time consumption. **Outdoor weathering is thus the most time-consuming type of accelerated testing.** If outdoor weathering were used only to test the coatings' ageing behaviour it would create a serious drag on the creation of new coatings. The reason for the low time-compressing effect is the fact that, in all cases, only a fraction of the day time is available for weathering. Figure 4.34 shows, first, the relative contribution of the daytime hours to the daily radiant exposure calculated for a cloudless 30th June (top), and second, the relative contribution of the month to the annual radiant exposure, as calculated for a fictitious cloudless year (bottom) for conventional outdoor weathering (5°) in South Florida.

As evident from the breadth of the white area in the upper diagram, the period from 6 a.m. to 6 p.m. contributes almost 100 % to the daily radiant exposure, as expected. Due to a lack of irradiation at night and the small amount of irradiance in the morning and evening hours (see Figure 3.26), some 50 % of the daily radiant exposure falls in the period from 10:30 a. m. to 01:30 p m., i.e. on one eighth of the day. Proposals to bridge the night hours by additionally using artificial radiation sources failed to produce the desired results. As may be seen from breadth of the white area in the bottom diagram, about 50 % of the annual radiant exposure occurs in the months of May, June, July, and August due to the low irradiance attainable during the winter season. This may be one of the reasons that outdoor weathering in Jacksonville, Florida reveals considerable damage only during the summer season, where high irradiance coincides with high moisture, high temperature, and acid atmospheric precipitation.

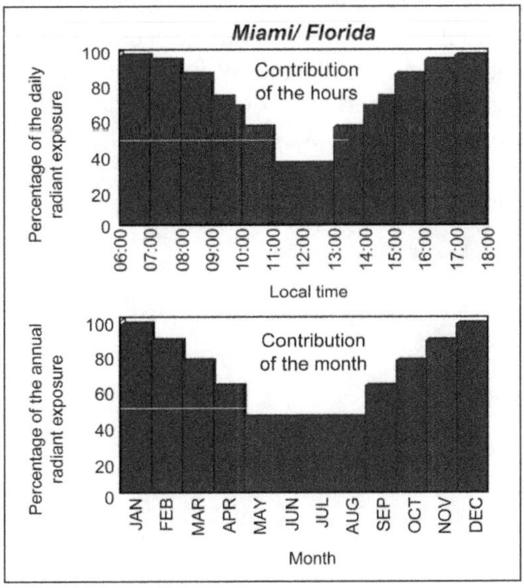

Figure 4.34: Contribution of the hours to the daily radiant exposure on 30th June (white area of the diagram on the top) and the month to the annual radiant exposure of a year (white area of the diagram on the bottom) for a clear sky only, for the reference climate Miami, Florida (adapted from [28])

As already shown in Chapter 3.26 clouds lead more or less to a decrease in radiant exposure. As will be discussed later in Chapter 5.3, even in the dry desert climate of Phoenix, Arizona, only about 66 % of the annual radiant exposure (direct portion of global radiation) maximally available in clear sky is available for concentration of radiation by optical means, due to clouds.

A further weakness of outdoor weathering is that a certain probability of failures must be expected, especially at weathering sites, where no permanent supervision by staff is possible. Extreme weather situations (storms, hurricanes), trespassing, theft, and vandalism, may lead to lost test panels or the kind of damage that makes assessing the weathering-related ageing behaviour impossible. As mentioned before, stopping an outdoor weathering test always means delaying coating development by one year, at least. **Note that outdoor weathering cannot be continued or restarted arbitrarily on any day of the year!**

A further weakness is the uncertainty of the weathering results caused by mostly unnoticed, unexpected, and unrecorded irregularities in the course of outdoor weathering. That is the case when other environmental stresses occur temporarily which are atypical of the weathering site, for instance, due to building work, heavier traffic, air pollutants, oil soot, tar, accidents in nearby industrial plants, and certain biological or chemical attack. Coatings cannot be expected to resist all possible forms of attack without any damage. The creators of such coatings often ask the question: "Was it poor coating quality or were irregularities in the respective outdoor weathering test responsible for the coating's premature failure?" On the other hand, typical weathering stresses sometimes do not occur during an outdoor weathering test. For instance, a temporary lack of acid atmospheric precipitation during outdoor weathering mostly goes unnoticed. Note that a good coating quality assessment based on such a benign outdoor weathering test may not be borne out during the coating's normal use.

Especially for European users of weathering services, the large distance from the recommended reference weathering sites must not be underestimated. Permanent or at least frequent supervision of their coatings under test is scarcely possible. Important information about the coating behaviour during weathering, such as how the coating damage occurred (e.g. spontaneously or gradually), reasons for a variety of damage (e.g. bird excrement), frequent falls from the weathering rack and the like mostly, is lost. Sensible decisions, for instance, about prematurely terminating the weathering test cannot be made at such a distance, without visual supervision by the coating specialist. As described in Chapter 4.3, key weathering stress factors can be controlled in special types of outdoor weathering. However, the variations in the annual radiant exposure (global radiation) remain a weakness of all types of outdoor weathering.

5 *Accelerated outdoor weathering by means of light concentration*

5.1 *Expectations*

The knowledge that sunlight can be concentrated by means of a concave mirror to the extent that it can ignite combustible materials, such as paper and fabric, led to proposals to exploit this effect for accelerating outdoor weathering. Increasing the irradiance by concentrating the sunlight meets a typical requirement for the accelerated form of accelerated testing, such as described in Chapter 2.5, in accordance with VDI Guideline 3958-01 [9]. Acceleration of coating ageing commensurate with the degree of light concentration was generally expected. Since the sun serves as radiation source, irradiation is free as it is in all other types of outdoor weathering. In other words, irradiation does not incur any energy costs. As also described in Chapter 2.5, all accelerated test methods employ one or more exceeded weathering stresses. However, the stress factors are for the most part substantially intensified relative to those found in "normal use".

Note that increased weathering stresses do not necessarily lead to a correspondingly increased ageing rate! Acceleration by exceeding the test parameters always means a serious interference with the coating surface micro climate and is no longer identical with that arising from the intended normal use. The particular significance of the coating surface micro climate is discussed in Chapter 3.2.5. Excessive acceleration-induced deviation from the worst case weather scenario of the coatings to be tested can even slow down the coating ageing. Furthermore, ageing processes may be triggered that never occur during normal use. However, the use of the sun as radiation source prevents the coatings' ageing behaviour from being influenced by irradiation of shorter wavelength than is found in reality. In other words, it is a certainty that the photons of concentrated solar radiation will never have a higher energy than that of the global radiation.

The discussions in this chapter are intended to keep the test engineer away from harbouring overly high expectations of acquiring information for predicting the coatings' service life more quickly by using light concentration than by conventional outdoor weathering. They are intended moreover to provide the test engineer with the necessary expertise to better decide whether the conventional or the accelerated method of outdoor weathering should be preferred for his special purpose. On the other hand, they are also designed to aid a more critical assessment of results gained by accelerated outdoor weathering and to provide a better explanation for deviations from normal use.

5.2 *Technical principles*

For the purpose of accelerated outdoor weathering, corresponding technical means of concentrating the solar radiation were developed in USA during the 1960s. The test engineer was provided with an international standard in the form of ISO 877, Part C: "Method of exposure to intensified weathering by daylight by means of Fresnel mirrors" [1]. This describes all the essential requirements for a device to perform outdoor weathering accelerated by concentrated solar radiation. The corresponding apparatus is properly called a Fresnel-reflecting concentrator in ISO 877. SUN 10, Q-TRACK, and EMMAQUA (acronym for Equatorial Mount with Mirrors for Acceleration with Aqua) are commercial names for such apparatus. Nowadays, Fresnel-reflecting concentrators can be found in high numbers at weathering sites, predominantly in central Arizona, and are run by well-known operators [52, 62]. Figure 5.1 shows a schematic diagram (left) and photograph (right) of a Fresnel-reflecting concentrator in accordance with ISO 887.

Ulrich Schulz: Accelerated Testing:
Nature and Artificial Weathering in the Coatings Industry

ISBN: 978-3-86630-908-1

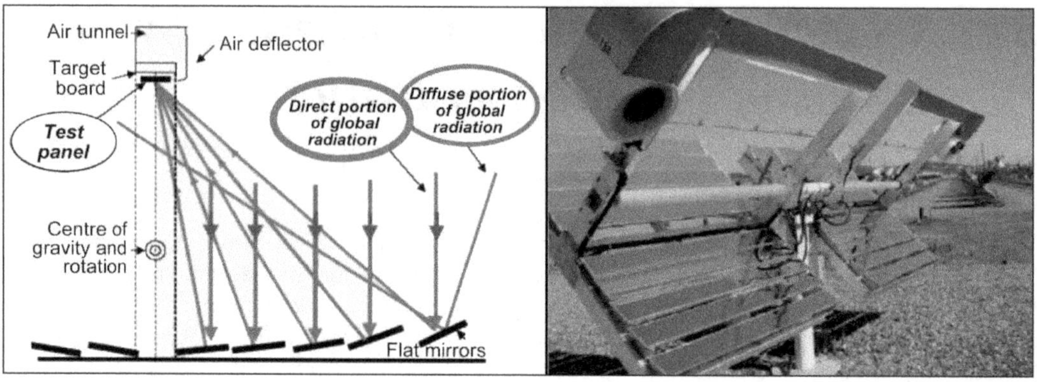

Figure 5.1: Schematic (left) and photograph (right) of a Fresnel-reflecting concentrator device in accordance with ISO 877, Method C

The solar radiation is concentrated by a system of 10 flat mirrors arranged to simulate tangents to a parabolic trough that reflects the sun's rays uniformly onto the target board. The test panels are mounted in the target area. EMMAQUA, for instance, uses mirrors measuring 15 cm x 142 cm that concentrate solar radiation on a target board measuring 13 cm x 140 cm. The test panels immobilized on the target board almost face the ground when the sun is in its azimuth. In order to reach maximum irradiance by directly incident solar radiation, modern apparatus closely tracks the path of the sun continuously through the day, in the same manner as the tracking rack discussed in Chapter 4.3.3.

A motorized sun-tracking mechanism moves the apparatus on two axes, making sure that the sun's rays always fall vertically on the test panels. Of course, the mirrors have to satisfy high requirements as regards surface evenness, reflectance, and cleanness in order that as much solar radiation as possible may be reflected. As has been mentioned several times in the preceding chapters, only a tiny portion of received solar radiation, the UV range with about 6 %, is needed for coating ageing in most cases. The bulk of global radiation (VIS and NIR) leads to heating of the test panels. This can produce high coating surface temperatures if readily absorbent coatings or coatings on poorly heat-conducting substrates (e.g. plastic coats) are exposed. In extreme cases, the coatings to be tested can be destroyed by melting. Fresnel-reflecting concentrators must be equipped with proper means for avoiding excessively high heat-curing stress.

Commonly, the test panel surfaces are subjected to strong cooling by means of heavy, controllable air circulation. The intensity of air cooling, ambient temperature (used for cooling), thickness, reflectance, absorbance, and heat-curing properties of the test panels determine their surface temperature. As reported in [52], it may be possible to limit the temperature to 10 °C above that of identical test panels weathered simultaneously by conventional methods at the same location. This can be achieved by making the cooling air flow directly along the test panel surface at a speed of at least 10 m/s.

Since the cooling air is directly taken from the ambient, i.e. usually from the dry desert environment, extremely dry air circulates over the test panel's surface and leads to extremely small relative humidity immediately on the coating surface. High temperature combined with low relative humidity inevitably leads to strong drying of the coating. The influence of the relative humidity during weathering cannot be tested. Weathering results can be expected which are not transferable to real practice, especially where moisture-sensitive coatings are tested. With the aim of correcting this unrealistic coating surface micro climate, modern Fresnel-reflecting concentrators are equipped with nozzles to spray oscillating deionised water onto the test panel surface. EMMAQUA is one of the Fresnel-reflecting concentrators predestined for this purpose. ISO 877 recommends appropriate cycles for wetting the test panel surfaces, during irradiation and at night.

To prevent dirt or salt crusts on the test panels or the mirrors, high requirements are imposed on the quality of the used water (for instance, <1 ppm particulate matter, <0.2 ppm silicate). Circulation of the water without continuous cleaning cannot be recommended. As already discussed in Chapter 4.3.6, the duration of the simultaneous action of water and UV radiation (TOW+U) is particularly important coatings testing. The sense or non-sense of special spray cycles recommended in ISO 877 for coatings is discussed in more detail later in Chapter 5.4.

5.3 Assessing the concentration of natural radiation

The Fresnel-reflecting concentrator type EMMAQUA, in accordance with ISO 877, equipped with 10 mirrors, and operated in Phoenix, Arizona, serves as a good example for assessing the degree of radiation concentration, because essential irradiation data are regularly published for this device only [30]. The following calculation demonstrates the efficiency of concentration to a good approximation.

First, let us assume for simplicity that the Fresnel mirror system reflects 100% of the incident solar radiation in the total range of the global radiation from 300 to 2500 nm, and continuously tracks the sun throughout the day in order to maintain the focus of the sun's rays on the test panel surface. In this case, the irradiance must be concentrated by the factor 10. Let us further assume that the tracking increases the annual radiant exposure by the factor 1.5 relative to that of conventional outdoor weathering performed simultaneously at the same location at a constant exposure angle (5°), as described in Chapter 4.3.4. Multiplied by the number of the mirrors, a 15-fold strengthening of the irradiation stress must theoretically be possible. In comparison to the data published by [30], EMMAQUA produces a significantly lower concentration of radiation in practice.

Figure 5.2 shows the values of monthly radiant exposure for the total range from 300 to 2500 nm (EMMAQUA-total), as well as for the UV range from 300 to 400 nm (EMMAQUA-UV) for 2004. The diagram also shows the concentration factors actually reached relative to conventional outdoor weathering performed simultaneously at the same weathering site at a constant exposure angle of 5°. The use of radiation data from conventional outdoor weathering tests performed at other exposure angles (34° or 45°) for comparison with the EMMAQUA returned roughly the same concentration factors as those for 5°. **In spite of a maximal 8-fold concentration in total solar radiation in June 2004 and a 5-fold concentration in UV radiation in October 2004, concentration factors of just 6.5 in terms of EMMAQUA-total and 3.7 in terms of EMMAQUA-UV were actually reached on average over a period of six years from 2000 to 2005, based on the data published monthly by the operator [30].**

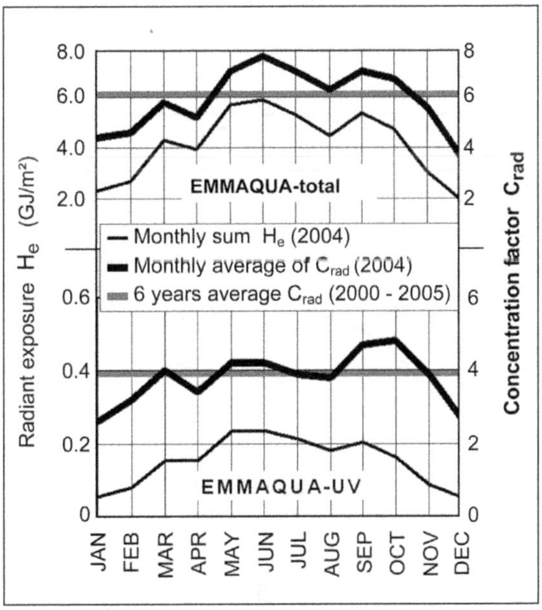

Figure 5.2: Monthly radiant exposure and the monthly average concentration factor (C_{rad}) and its average from January 2000 to December 2005 for the total wavelength range 300 to 2500 nm (EMMAQUA-total in the upper diagram), as well as for the UV range 300 to 400 nm (EMMAQUA-UV in the bottom diagram) attained by means of a EMMAQUA Fresnel-reflecting concentrator operated in Phoenix, Arizona (adapted from [30])

This unexpectedly low concentration factor raises the question: "What are the reasons

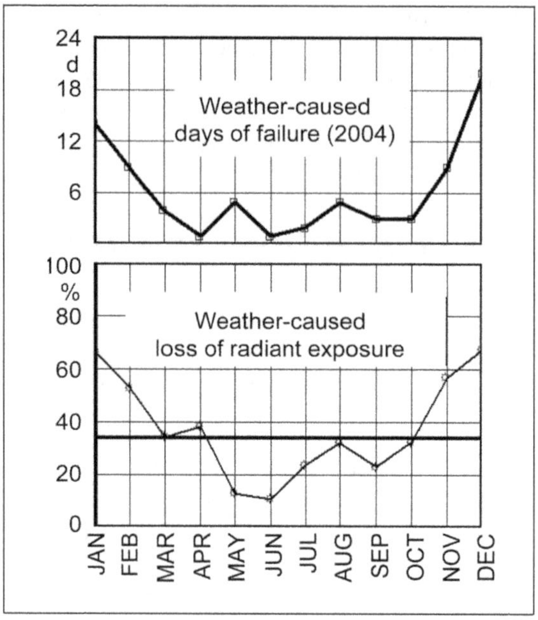

Figure 5.3: Weather-induced failure days in which no concentration of radiation was possible (upper diagram) and the overall weather-induced loss of radiant exposure (bottom diagram) arising from the use of an EMMAQUA Fresnel-reflecting concentrator operated in Phoenix, Arizona

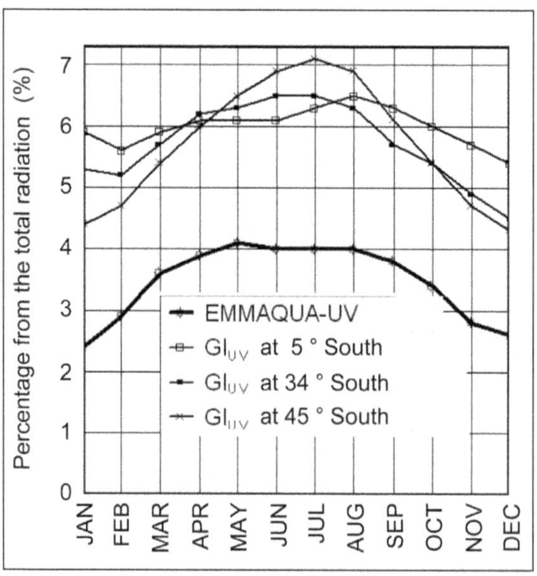

Figure 5.4: Percentage of UV (EMMAQUA-UV) from the total radiation (EMMAQUA-total) concentrated by means of an EMMAQUA Fresnel-reflecting concentrator operated in Phoenix, Arizona, compared with the percentage of global UV ($Gl_{UV\text{-}x}$) from the global radiation (non-concentrated) measured at three different exposure angles for the months of the year 2004 (adapted from [30])

for the unexpectedly low efficiency of the EMMAQUA in terms of the average radiant exposure, especially in the UV range of the solar radiation?" The fact that the **reflectance of the mirrors** of the Fresnel-reflecting concentrator never reaches 100 % is the first reason. As for optical quality, ISO 877 recommends only the use of mirrors with a spectral reflectance of at least 65 % at 310 nm, obviously out of consideration for the economic aspects of producing better-reflecting materials. Furthermore, ISO 877 recommends that the mirrors be washed daily at least in order that a high level of reflectance be maintained. However, the question remains: "Is daily washing enough, especially in highly dusty desert areas?" The weather is the second reason. Although it is recommended as a hot-dry reference climate, the weather of Phoenix, Arizona, is not optimal for operating a Fresnel-reflecting concentrator. For instance, 76 days could not be used for concentrating in 2004, because of a lack of direct solar radiation, obviously caused by clouds! Since the test panels predominantly face away from the sky, the contribution which the diffuse portion of global radiation impinging on the test panel surface makes to irradiation stress can be neglected.

Figure 5.3 shows, again for the year 2004, how the weather influence is distributed over the year. The upper diagram shows the number of days without direct solar radiation. The lower diagram shows the total weather-induced loss of direct radiation. The highest daily radiant exposure for the respective months, multiplied by the number of days, was taken to be the 100 %-value for the calculation. As can be seen, the loss of direct radiation is concentrated in the winter months. The figure reveals that a third of the hours which could have been available under clear skies to operate EMMAQUA in 2004 were unsuitable. **Regions with high cloud cover, such as Florida and central Europe, are unsuitable for operating Fresnel-reflecting concentrators. A suitable location should have at least 3500 hours of sunshine a year.**

The fact, that **only direct radiation can be concentrated by optical means**, such as mirrors, is the third and crucial reason for the relatively low concentration factors, especially in UV. The diffuse portion of the global radiation (sometimes called diffuse sky radiation) cannot be used by Fresnel-reflecting concentrators. As discussed in more detail in Chapter 3.2.1.3, the diffuse portion increases and the direct portion decreases with decreasing angle of elevation of the sun. As a result, the increase in irradiance reached by tracking the sun throughout the day is almost lost due to the decrease in the direct portion in the course of the day.

As also discussed in more detail in Chapter 3.2.1.3, the extent of scattering, as a cause of the diffuse radiation, depends heavily on the wavelength. The part of the solar radiation that lends itself to concentration by a Fresnel-reflecting concentrator decreases with decreasing wavelength. UV is scattered much more than VIS and NIR. Consequently, that portion of global UV which lends itself to concentration decreases particularly strongly with decreasing angle of elevation of the sun. This manifests itself in the fact that the radiation concentrated by EMMAQUA contains less UV than the global radiation.

Figure 5.4 shows the extent to which the ratio of UV (300 to 400 nm) to total radiation (300 to 2500 nm) changes due to radiation concentration by means of EMMAQUA. In this figure, the portion of UV as a percentage of the total radiation additionally changes from month to month. The influence of scattering can easily be seen from the particularly low percentage of UV in the winter months. As can be seen, global radiation contains about 6 % UV on average, almost irrespective of the exposure angle. In contrast, the concentrated solar radiation (EMMAQUA-total) contains only 3.5 % UV on average (2000 to 2005). From the discussions in Chapter 3.2.1.2 of the activation spectrum, UV radiation at and near 320 nm contributes most to coating ageing. Data for the wavelength range of 320 ± 0.5 nm are not published by the operator of the weathering site. Therefore, it seems logical to calculate the data for 320 ± 0.5 nm by means of CESORA, with a view to assessing the efficiency of a Fresnel-reflecting concentrators for this specific narrow wavelength range. The data were calculated for Phoenix, Arizona. Figure 5.5 shows the calculated values of the diffuse and direct portions as a percentage of the sum of both portions (whole natural irradiation at 320 nm) as a function of the month of the year. It also shows the annual average value of the direct portion as a thick grey line parallel to the x-axis. As expected, the ratio of direct to diffuse radiation varies with the sun's angle of elevation. The curves in Figure 5.5 were calculated for a permanently clear sky.

In other words, even under a clear sky, only about a third of the total natural radiation available at 320 nm can be concentrated by a Fresnel-reflecting concentrator. Assuming that the reflectance of the mirrors is 70 % at 320 nm (in accordance with ISO 778), given the positive effect of tracking, and with allowance for the weather-induced loss of direct radiation found in 2004, the annual average irradiance or the annual radiant exposure at 320 nm only doubles relative to conventional outdoor weathering performed simultaneously at the same weathering site.

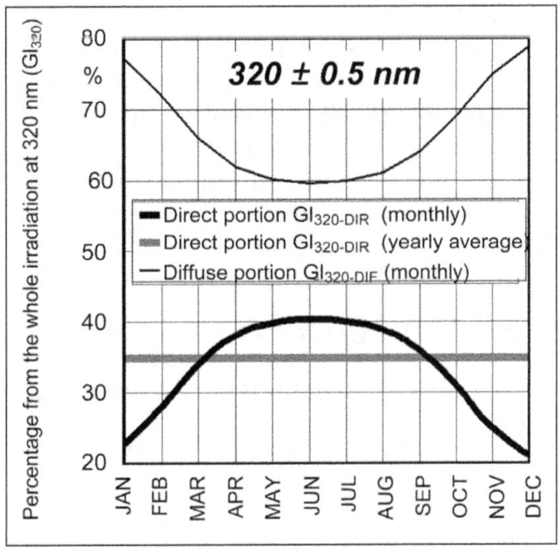

Figure 5.5: Direct und diffuse portion of global UV in the narrow wavelength range of 320 ± 0.5nm calculated for each month of a year for Phoenix, Arizona by means of CESORA [28]

Table 5.1: Annual radiant exposure actually attainable on the target board of an EMMAQUA Fresnel-reflecting concentrator, in accordance with ISO 877, Method C (operated in Phoenix, Arizona) compared with that attainable in conventional outdoor weathering in Phoenix, Arizona, and Miami, Florida, along with the respective concentration factors as a function of the wavelength range under consideration

Concentrating of solar radiation using EMMAQUA		Phoenix-EMMAQUA	Phoenix reference climate (34° N)	Miami-reference climate (26° N)
1		2	3	4
Total wavelength range 300 to 2500 nm	annual radiant exposure GJ/m²/a	51.0	8.4	6.6
	concentration factor Crad		6.5	8
UV 300 to 400 nm	annual radiant exposure GJ/m²/a	1.63	0.43	0.38
	concentration factor Crad		4	4
UV320 (320 ± 0,5)nm	annual radiant exposure (calculated) kJ/m²/nm/a	7.5	3.1	3.1
	Concentration factor Crad		2	2

Table 5.1 shows, subdivided into the three most important spectral ranges, the annual radiant exposure and respective concentration factors attainable by EMMAQUA Fresnel-reflecting concentrators operated in Phoenix, Arizona, relative to that of conventional outdoor weathering performed simultaneously at the same weathering site and in Miami, Florida. Although the aforementioned comparisons were done only for Phoenix, Arizona, it seems reasonable to include the Miami, Florida reference climate in the Table. For the total solar radiation from 300 to 2500 nm, the EMMAQUA achieved a concentration factor of about 6.5 relative to conventional outdoor weathering in Phoenix, Arizona, and a concentration factor of about 8 relative to conventional outdoor weathering in Miami, Florida.

As regards the UV radiation of 300 to 400 nm, identical concentration factors of about 4 were found, relative to the two aforementioned sites used for conventional weathering. The same statement holds for the narrow wavelength range of 320 ± 0.5 nm, for which identical concentration factors of just about 2 are obtained. The low-radiation winter months in Phoenix, Arizona, as shown in Figure 5.3, can be dispensed with in principle. **With regard to testing of modern automotive coatings, the commonly required two-year conventional outdoor weathering in Phoenix, Arizona, as well as Miami, Florida, produces at 320 ± 0.5 nm the same radiant exposure as is attained by a Fresnel-reflecting concentrator in central Arizona in the eight months from 1st March to 30th October.** Note that all the discussions here in Chapter 5.3 pertain only to the concentrating effect on solar radiation, and not to any effects relating to acceleration of coating ageing!

5.4 Concentration of natural radiation and acceleration of coating ageing

After discussing the concentration of natural radiation by means of Fresnel-reflecting concentrators, the question remains as to the level of attainable acceleration of coating ageing. The discussions below are intended to show that the radiation concentration factor cannot simply be viewed as identical with the acceleration factor for coating ageing. As mentioned earlier, coatings are often subdivided into moisture-sensitive and heat-sensitive types.

Concentrating the natural radiation alters not only the irradiation stress but also the other weathering stresses, such as the moisture and temperature of the coating surface micro climate.

Coatings differing in the aforementioned properties are expected to reveal different acceleration factors in spite of the same concentration factor. Studies are described in [63] in which coatings that differed extensively in their chemistry were simultaneously exposed on EMMAQUA in Phoenix, Arizona, and to conventional outdoor weathering in Miami, Florida. With 50 % of the loss of gloss serving as ageing criterion, acceleration factors of 5.8 (best) and 3.4 (worst) were achieved. Comparing these results with the curves in Figure 5.2, the average acceleration factor of 4.3 seems to be almost identical with the concentration factor of the natural UV radiation in this example.

Laboratory studies, described in [64], were devoted to the relationship between UV irradiance and the rate of photooxidation. Films about 10 µm thick of an acrylate melamine based clearcoat were used for this study. The specimens were irradiated only with UV radiation (300 to 400 nm) generated by an artificial radiation source. The study revealed linear proportionality between the concentration factor and acceleration factor if the increase in the oxidation products was tracked. This result held true up to a four-fold increase in irradiance. Since the specimens were irradiated only with UV radiation, the coating surface micro climate, in terms of temperature and moisture, could be seen as almost unchanged in the chosen range of concentration. This explains the good agreement that was found. In this case, it could be stated that UV radiation was predominantly responsible for the accelerated coating ageing, because enhanced heat-curing stress could be ruled out. However, this statement does not hold for concentrating natural radiation by means of Fresnel-reflecting concentrators. Here, the total spectrum of the natural radiation is involved in concentrating. The 6.5-fold increased irradiance, on average, can produce to an equivalent increase in heat-curing stress. The left diagram of Figure 5.6 shows the attainable temperature of specimens weathered on EMMAQUA, as a function of their radiation absorbance (left diagram) and canonical cooling. It proves that black panels can reach a surface temperature of 90 °C in spite of intensive cooling. The test panels were much more thermally stressed on the EMMAQUA than during conventional outdoor weathering and normal use alike. The right diagram of Figure 5.6 shows how fast the temperature increases if cooling of the test panels is stopped. Temperatures up to 150 °C can be reached on highly absorbent coating surfaces in a few minutes.

As described in Chapter 3.2.5, increasing surface temperature can lead to both acceleration and deceleration of coating ageing under the conditions of the natural environment. Therefore, it cannot be ruled out that the ageing effects of exposure on EMMAQUA, for instance, in the case of highly UV-stabilised coatings, are caused only by heat-curing stress and that the resultant acceleration factor randomly correlates with EMMAQUA's concentration factor.

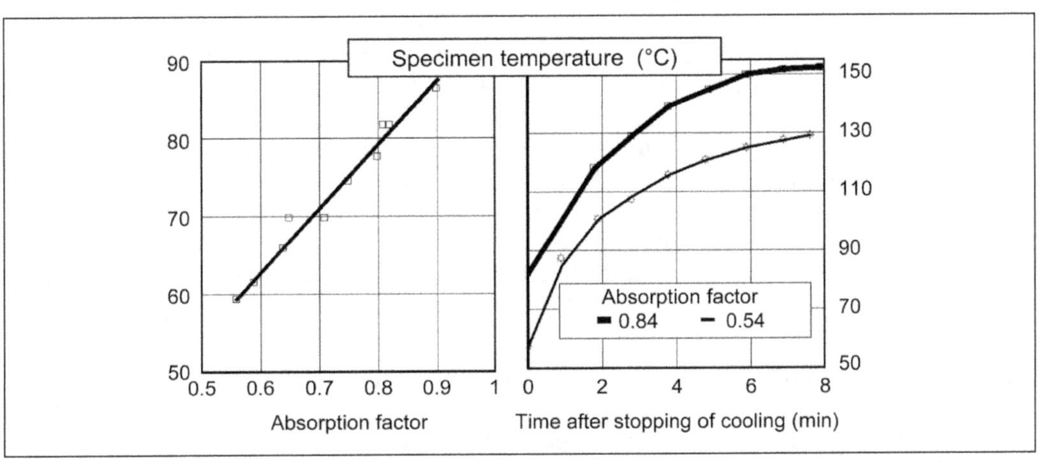

Figure 5.6: Surface temperatures of specimens of different radiation absorbency on an EMMAQUA Fresnel-reflecting concentrator (left), and their rapid raise after cooling has ceased (right), adapted from [12]

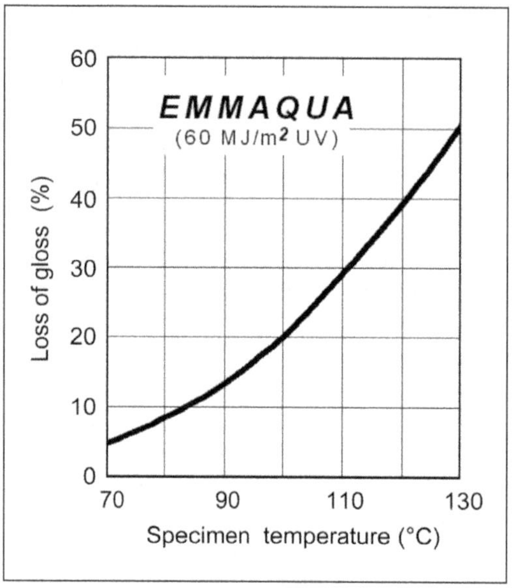

Figure 5.7: Loss of gloss by a coil coating after exposure on an EMMAQUA Fresnel-reflecting concentrator as a function of the surface temperature, based on identical UV radiant exposure (adapted from [54])

As also discussed earlier, unrealistically exceeded surface temperatures can initiate degradation processes that never occur in normal use. If coating ageing is caused only by heat-curing stress, coatings of different properties, exhibit ageing behaviour on EMMAQUA which contradicts that found in normal use. In such cases, coatings for weatherability comparisons can show a ranking after weathering on EMMAQUA which is the total opposite of that found in practice.

There is always the possibility that the best one on EMMAQUA is the worst one in normal use. For instance, a decision to the effect that the coating being is the best for the intended normal use may be wrong. The literature contains many references to EMMAQUA's correlation with practice. Some authors report good ranking correlation, and others poor correlation, between EMMAQUA and conventional outdoor weathering. Other authors [63], for instance, report poor ranking correlation in terms of loss of gloss (PEARSON = 0.68), yet good ranking correlation in terms of change in colour (PEARSON = 0.9). The PEARSON ranking correlation coefficient is predominantly used in coatings testing, as will be described in more detail later in Chapter 7.3. In the case of automotive coatings, poor ranking correlation was reported when the results of EMMAQUA weathering and conventional outdoor weathering were compared. Automotive coatings are usually cured at a minimum of 150 °C and therefore are mostly highly stabilised against heat-curing stress, too. Consequently, the temperature cannot be expected to fully unfold its potential to accelerate photochemical ageing on EMMAQUA, in accordance with Arrhenius' Law. Studies from [54] confirm this assumption for the example of coil coating, in which the test panel temperature was varied by changing the intensity of cooling.

Figure 5.7 shows the results of this study. The loss of gloss after identical radiant exposure is presented as a function of the absorbance of the test panel surface. The expected two-fold ageing rate per temperature increase of 10 °C failed to materialize. The reasons for the deviation from Arrhenius' Law may be the decrease in the most effective UV (especially at 320 nm), as well as a marked change in the coating surface micro climate, both due to the concentrating of natural radiation. As described in Chapter 3.2.5, an increase in surface temperature inevitably leads to a decrease in the relative humidity directly on the coating surface under the conditions of the natural environment. A radiation-induced increase in surface temperature, as well as the use of extremely dry air taken from the desert environment for cooling the test panels, leads to extreme and highly penetrative drying of the coating layer. This can be seen as the key reason for the poor correlation between EMMAQUA's weathering results and practice, especially in the testing of automotive coatings, whose ageing behaviour is known to vary markedly as a function of the coating's water content during irradiation.

Improving the correlation with practice might be the aim behind cyclical wetting, which is offered as an option for weathering on EMMAQUA, and is expressed in the acronym by the suffix "AQUA". As already discussed in more detail in Chapter 4.3.6, methods of wetting commonly practiced and standardised by ISO 877 recommend cyclical wetting, mostly overly short spray phases or overly long dry phases. Using these recommendations can be expected to decrease the intensifying effect of simultaneously acting UV and water to insignificance. In other words, spray phases in accordance with ISO 877, for instance, cycle No. 1 which comprises 8 minutes' water spraying and 52 minutes' dry conditioning, both during irradiation, plus three night-time sprays, cannot be assessed as a means of triggering the accelerating effect of the mutual action of water and UV significantly. Figure 5.8 shows the change in BST during exposure on EMMAQUA by means of spray cycle No. 1 (ISO 877) over an exposure period of 4 days. The black standard thermometer was mounted on the target board, and therefore also subjected to water spraying. The 8-minute spray phases during the hours of daylight can be clearly seen. The oscillating spraying caused abrupt changes in surface temperature of almost 30 °C (from about 30 °C to about 60 °C). However, the recommended spray phases were always too short, for instance, to saturate a 30 µm-thick clearcoat with water, as shown in Figure 3.41.

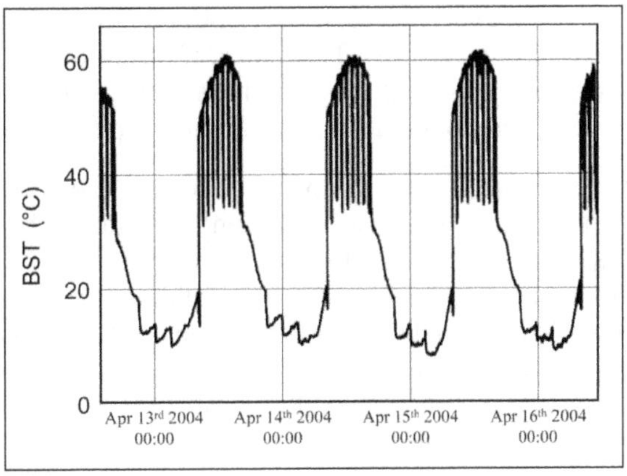

Figure 5.8: Black standard temperature (BST) measured on the target board of a EMMAQUA Fresnel-reflecting concentrator by means of a black standard thermometer over several days' exposure involving phases of cyclical wetting by artificial rain (adapted from [41])

The water which had penetrated into the polymer left the coating layer rather fast after spraying ceased, because the surface temperature always reached its respective maximum again very rapidly. The relatively high test panel temperature and the low water content of the cooling air accelerate the rate of diffusion of the water from the polymer. Therefore, only the spraying phases (8 minutes a cycle) can be considered available to the TOW+U (simultaneous action of water and UV) for the worst case weather scenario for weathering of moisture-sensitive coatings.

Cycle No. 4 (ISO 877), for instance, which recommends spraying only at night is nonsense where coatings testing is concerned, because the most effective TOW+U is not considered at all. Abrupt changes in surface temperature, as can be seen in Figure 5.8, too, and the water content in the coating layer are often seen as accelerating coating ageing. Changing mechanical stress caused by changing temperature or water content is also generally considered to be part of weathering stress. As already discussed in more detail in Chapters 3.2.2 and 3.2.3, more recent experience shows that the aforementioned changes make it possible only to recognise the state of coating ageing earlier, but do not contribute to photochemical ageing.

Since the weathering phases on highly dried coatings greatly outnumber those on wet coating surfaces in all of the aforementioned cycles of ISO 877, the accelerating effect of the EMMAQUA on the coating ageing can fail, especially in the case of moisture-sensitive coatings, despite multiply concentrated irradiance. In this case, there is a possibility that the conventional outdoor weathering in a hot-humid climate, like Miami, Florida, leads to faster coating ageing than exposure on EMMAQUA in Phoenix, Arizona. This result can be expected with special types of modern automotive coatings which are equipped with highly effective paint additives for stabilising simultaneously against irradiation and heat. Such a case was reported in [103].

Two identical series of 27 black-coloured basecoat/clearcoat systems for automotive OEM, equipped with different water-borne two-component polyurethane clearcoats, were weathered in parallel on EMMAQUA in Phoenix, Arizona, and on the black box in Miami, Florida in 2002. In this example, the result of one year of exposure was that 26 of the 27 tested coatings showed considerably higher loss of gloss after exposure on the black box than after exposure on the EMMAQUA in spite of roughly 4-fold UV radiant exposure. A PEARSON of 0.08 provided absolutely no ranking correlation. Fortunately, ISO 877 permits the option of other spray cycles if this is agreed between the interested parties. **Therefore, if automotive coatings are to be tested, sensible cycles of long wet phases during daylight hours should be agreed with a view to attaining greater acceleration of coating ageing and better correlation with practice.**

In order to unfold the full potential to enhance the efficiency of the EMMAQUA, it is recommended that the test panels be permanently irradiated behind a UV-transparent window or at times in a special container with flowing water in accordance with [66]. This kind of mounting leads to complete isolation from ambient climatic conditions. A combination of high TOW+U, roughly fourfold-exceeded UV-irradiance, and a not unrealistically exceeded surface temperature facilitate weathering of the test panels on EMMAQUA for longer under the conditions of the worst case weather scenario, especially for moisture-sensitive coatings. Higher accelerating factors and correlation with practice can be expected.

Reliable predictions about a coating's service life on the basis of weathering on Fresnel-reflecting concentrators can only be made if the weathering results correlate with the conventional outdoor weathering on the favoured weathering site, for instance, Miami, Florida. If identical series of different coatings are to be tested, different rankings of the coatings in terms of their weatherability must always be reckoned with. Weathering of new coating creations by means of Fresnel-reflecting concentrators as the first or only weathering method practiced cannot be recommended. **Note that agreement between rankings for the ageing rate of coatings weathered conventionally and weathered on a Fresnel-reflecting concentrator is scarcely likely because of the diversity of modern coatings.**

Supervision of series production is recommended if the acceleration factor is determined and the type of coating ageing proves to be more or less identical with that occurring in normal use. With regard to accelerated testing by means of Fresnel-reflecting concentrators, it may be said in summary that the concentration of natural radiation and acceleration of photochemical coating ageing are not equivalent. **Since the coating surface micro climate deviates substantially from that which forms in normal use, identical concentration factors can lead to markedly different acceleration factors for coating ageing.**

5.5 Assessment of repeatability

As with each kind of accelerated testing, a distinction has to be made between the repeatability of weathering stress and the repeatability of coating ageing. First, let us look at the repeatability of the irradiation stress during weathering accelerated by concentration of natural radiation. Figure 5.9 shows the annual radiant exposure of the EMMAQUA total (300 to 2500 nm) and EMMAQUA-UV (300 to 400 nm) published by [30] for a period of eight years from January 2000 to December 2007. EMMAQUA-total and EMMAQUA-UV vary by about 22 %.

Compared with the curves in Figure 3.27, the annual irradiation stress in the case of weathering on the EMMAQUA fluctuates roughly twice as much as in conventional outdoor weathering in Phoenix, Arizona, and Miami, Florida. This raises the question as to "what is the reason for the higher variance?" Like conventional outdoor weathering, the weathering stress factor irradiation acting on Fresnel-reflecting concentrators is also coupled to the natural radiation. However, both the direct and diffuse portion of global radiation contribute to irradiation stress in conventional outdoor weathering. In contrast, EMMAQUA can only use the direct portion for concentrating the radiation. As described in Chapter 3.2.1.4, clouds acting as randomly occurring parts of the weather were established as the main reason for the fluctuation in global radiation. As explained there, the direct portion of the global radiation decreases with increasing clouds much faster than the diffuse portion. However, diffuse radiation cannot be concentrated. Therefore, identical fluctuations in the degree of clouds influence the irradiance on the EMMAQUA much more than is the case for conventional outdoor weathering. In other words, the **repeatability of radiation stress during weathering on Fresnel-reflecting concentrators is worse than is the case for conventional outdoor weathering.**

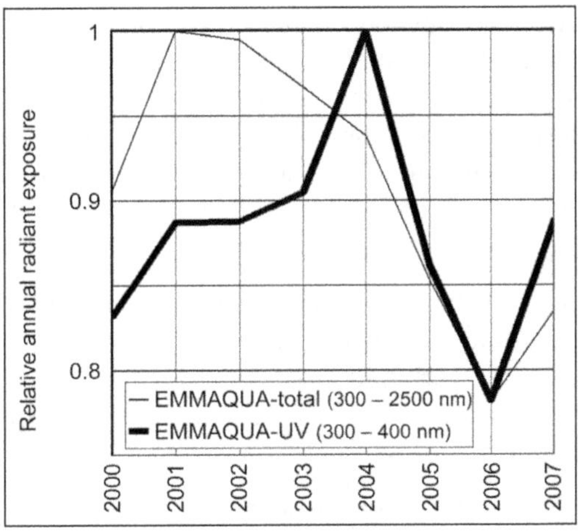

Figure 5.9: Relative annual radiant exposure (total und UV) published for EMMAQUA Fresnel-reflecting concentrators in Phoenix, Arizona, for a 7-year period from 2000 to 2007 (adapted from [30])

Now, let us look at the other weathering stresses, such as heat and water. As is known, the poor repeatability of conventional outdoor weathering is due, in addition to the varying global radiation, in large part to the randomly varying temperature and relative humidity.

In accelerated outdoor weathering by means of natural radiation concentration, only the weather episodes with direct radiation are important. By that is meant the time when the sun's rays fall, unhindered by clouds, on the test panel surface. During such weather episodes, the coating surface micro climate, in terms of temperature and relative humidity, is largely isolated from the ambient climate. The direct portion of the global radiation arriving at the EMMAQUA's mirrors remains the only link to the ambient climate. The temperature and relative humidity on the coating surface are essentially influenced by the irradiance of the concentrated natural radiation and the human-controlled intensity of the air cooling system. If the change in the surface temperature is completely human-controlled by a fixed programme during the day, the heat-curing stress is perfectly repeatable. Influences due to fluctuation in the relative humidity through the use of dry ambient air cannot be expected. The only essential water stress is the cyclical wetting of the test panel surface, and is human-controlled, too. On account of the extensive isolation of the coating surface micro climate from the ambient climate, better repeatability of the heat-curing stress and the stress water can be expected for weathering on Fresnel-reflecting concentrators than for conventional outdoor weathering. This can be assessed as an essential advantage of Fresnel-reflecting concentrators. In summary, the discussions have shown that, **in accelerated testing by means of EMMAQUA Fresnel-reflecting concentrators, the repeatability of the irradiation stress is worse, but the repeatability of the heat-curing stress and the water stress is better than is the case for conventional outdoor weathering.**

Now that the repeatability of weathering stress has been dealt with, the question about the repeatability of coating ageing remains. Sufficiently reliable publications that answer this question are nowhere to be found. Since the influence of natural fluctuation in ambient temperature and moisture is largely suppressed, the repeatability of EMMAQUA's weathering results cannot be expected to be worse than that of conventional outdoor weathering, in spite of the higher fluctuation in irradiance.

5.6 *Advantages and disadvantages of accelerated outdoor weathering*

As is the case for all other methods of outdoor weathering, the **first advantage of accelerated outdoor weathering is the use of the sun as radiation source**. This rules out the possibility that the test panels will be attacked by UV radiation of wavelength shorter than the natural cut-on of 300 nm. As already discussed in Chapter 4, radiation whose wavelength is shorter than that of the global radiation, can cause ageing processes that never occur in normal use. The irradiation itself is free. Therefore, despite the greater technical expenses, the costs of a weathering service by means of Fresnel-reflecting concentrators, expressed in terms of a square meter test panel surface, are still lower than those of laboratory weathering with artificial radiation sources. The **second advantage of accelerated outdoor weathering is multiply enhanced irradiance** acting on the test panel surface, although the original expectations in terms of the degree of concentrating of UV have not been fulfilled.

The **good repeatability of the heat-curing stress and the water stress can be seen as a third advantage.** The reason for the poor repeatability of the weathering stresses in conventional outdoor weathering has largely been eliminated. This is particularly important because the surface temperature and the coatings' water content are known to have a pronounced influence on coating ageing. Better repeatability of the weathering results can be expected.

Now we come to the disadvantages and weaknesses of accelerated outdoor weathering. The **greatest weakness may be the fact that the coating surface micro climate deviates strongly from that of normal use**, as described in more detail in Chapter 5.4. Different acceleration factors must always be expected when different coatings are tested. This means that the acceleration factor depends heavily on the type, formulation and application of the coatings to be tested. Therefore, coatings in test series will mostly yield weatherability rankings which disagree with those found in conventional outdoor weathering and normal use alike. Since the sun serves as radiation source, accelerated outdoor weathering is disadvantaged by the limited hours of sunshine and the low irradiance during the morning and evening hours, as is the case for other weathering test performed outdoors, too. **Accelerated outdoor weathering cannot make full use of the available solar radiation because only its direct portion can be concentrated by optical means.** In contrast, conventional outdoor weathering also uses the diffuse radiation. That can be seen as a further major drawback of radiation concentration.

A **third disadvantage is the altered spectral distribution of the radiation** acting on the test panel surface, caused by concentration of the solar radiation. Since radiation scattering varies with the wavelength, less UV radiation is available for concentrating. Unlike conventional outdoor weathering, only about half of the UV irradiance can be found on the target board of a Fresnel-reflecting concentrator. The test panels are much less subjected to the most ageing-effective part of solar radiation around and below 320 nm. Since diffuse radiation cannot be concentrated, the **repeatability of the irradiation stress during weathering on Fresnel-reflecting concentrators is worse than in the case of conventional outdoor weathering**, as already discussed in Chapter 5.5. In contrast, highly repeatable heat-curing stress and water stress were mentioned as one of the advantages. Recall that the surface temperature and wetting of the test panels are commonly controlled by the intensity of cooling and immobilized cycles of spraying by water. However, the breadth of that control is severely limited to a relatively high surface temperature during the hours of highest irradiance only.

Where there is compliance with ISO 877 as regards the frequency and duration of spraying cycles, there is too little water stress relative to that of conventional outdoor weathering in Miami, Florida, and normal use. In other words, coating weathering on the EMMAQUA in the conventional manner is performed under overly dry weathering conditions. It must therefore be assumed that the good repeatability of the heat and water stresses is diminished by the imposition of climatic test conditions that do not occur in normal use, as well as only weakly contributing to coating ageing.

This discrepancy between the climatic conditions for weathering on the EMMAQUA and climatic conditions in normal use can be seen as a disadvantage, too. For instance, the coating surface micro climate, namely 37 °C/>80 % relative humidity, which frequently occurs in Miami, Florida, in the summer, combined with the global UV available between rain showers, cannot be imitated on EMMAQUA's target board. As experienced with automotive coatings, precisely this coating surface micro climate can meet the worst case weather scenario, especially for moisture-sensitive coating types. In other words, the hours which meet the worst case weather scenario during conventional outdoor weathering can yield rates of coating ageing that will never be reached on the EMMAQUA at any time.

Balancing the advantages and disadvantages, it can be stated: If the acceleration factor of coating ageing is shown by appropriate preliminary studies to be economically justifiable and sufficiently repeatable, outdoor weathering on Fresnel-reflecting concentrators can be a suitable, effective type of accelerated testing for monitoring series production for certain coatings or coating systems. **However, tests on new coating creations will inevitably lead to false service life predictions.**

6 Artificial weathering - accelerated testing in the laboratory

6.1 Aim of artificial weathering

Overcoming the weaknesses of outdoor weathering without dispensing with its strengths is the general aim of artificial weathering. The weakness of being highly time consuming can be relatively easy overcome because all periods that make little or no contribution outdoors to coating ageing can be excluded in artificial weathering. It may be said of modern coatings that that no photochemical ageing takes place if irradiation is absent. Therefore, the dark phases of testing are no longer necessary in coatings testing. Nevertheless, dark phases are still seen as sensible for certain polymers and are still included in some standardized weathering test methods. The possibility of continuously irradiating the test panels is one of the essential aspects of the time-saving potential of artificial weathering. The weather episodes that most contribute to coating ageing, namely the worst case weather scenarios, as described in Chapter 2.5, occur with limited frequency outdoors. In order to gain optimal time-compressing effects, artificial weathering focuses on the worst case weather scenario(s) of the coatings to be tested. Subjecting the test panels to their worst case weather scenarios more often than occurs outdoors, or continuously if only one worst case weather scenario exists, is the best way to achieve maximum time compression. In order to achieve an accelerating effect as well, the artificial weathering additionally focuses on exceeding individual weathering stresses, such as irradiation, heat, or water during the worst case weather scenario. Furthermore, artificial weathering has the potential to avoid randomly occurring deviations from the fixed test parameters.

Uncontrollable deviations in weathering stress from test to test are named in Chapter 5.6 as one of the key weaknesses of conventional outdoor weathering. If performed correctly in correctly working laboratory devices, artificial weathering has the potential to also achieve much better repeatability and reproducibility than conventional outdoor weathering. As commonly used in artificial weathering, repeatability describes the ability of achieving identical weathering results, when the same weathering test is repeated on the same kind of test panels in the same laboratory weathering device. Reproducibility refers to the ability to gain identical results, when repeated in a different laboratory weathering device operated in the same or a different testing laboratory.

Theoretically, the state of the art should make exact repetition of weathering stresses in laboratory possible. However, good repeatability and reproducibility necessitate powerful laboratory weathering devices, which are relatively expensive. In other words, technical and economic outlay often imposes limits on purchasing and operating. **However, it should be noted that the cheaper the weathering device, the worse is the repeatability of the weathering results!**

As described in Chapter 4.3, improving the repeatability by controlling certain weathering stresses, such as heat, water, and acids, has already been tried in special outdoor weathering methods. However, the annual fluctuation in irradiation remained uncontrollable, because natural radiation was used. This source of uncertainty in outdoor weathering results can only be overcome by the use of artificial radiation sources that can be seen as the most essential feature of artificial weathering.

As mentioned in Chapter 5.6, closeness to reality is one of the advantages of conventional outdoor weathering. Good simulation of natural weathering stresses, especially of those revealing synergies, mostly requires high technical outlay. However, **it is generally considered impossible and often not meaningful to imitate Nature completely and exactly. However, it should be**

Ulrich Schulz: Accelerated Testing:
Nature and Artificial Weathering in the Coatings Industry

ISBN: 978-3-86630-908-1

noted that incorrect or incomplete simulation of those natural weathering stresses which occur in normal use can lead to false predictions of weatherability and service life for the coatings to be tested! Therefore, the attainment of good correlation with practice and adequate repeatability imposes high requirements on testing technique, test method, and care on the part of the staff involved in coatings testing.

6.2 Simulation of natural weathering stresses in the laboratory

6.2.1 General aspects

Modern test practice is predominantly oriented towards standardized weathering test methods as regards the choice of artificial radiation sources and the climatic test parameters. As a result, the available potential for time compression or acceleration is often not optimally used, mostly because of insufficient knowledge on the part of the staff involved in coatings testing as to the respective worst case weather scenarios for the coatings to be tested. However, the design of new, or the extension of existing, weathering test standards cannot keep pace with developments in new coating types, or new experiences in coatings testing. As described in more detail in Chapter 8, standardizing takes too much time for it always to be up to date. Conducting coating tests on the basis of different weathering test methods inevitably leads to different predictions about ageing behaviour in normal use. Comparisons of the ageing behaviour of coatings tested in series would create the impression that different methods of artificial weathering would yield rankings different from those yielded by outdoor weathering. Ranking correlation with outdoor weathering or, better, normal use serves as a general measure of how correctly an artificial weathering test method simulates the natural weathering stresses.

There are many publications that allow the results of artificial weathering to be compared with those of outdoor weathering [16, 22, 67–70, 103, 104]. Those authors report more or less good correlation. The studies described by [16] and [22] have been chosen for discussions below. Their results can be seen as representative, trustworthy and sufficiently up to date to serve as proper examples. The standardized weathering test methods used therein are in widespread use in coatings testing at present. Their essential features are listed for each author in Tables 6.1 and 6.2.

A look at the weathering methods listed in the first lines of the tables reveals great diversity. They differ in the type of artificial radiation source. As described in more detail later in Chapter 6.2.2.1, xenon lamps (denoted X in the acronym for the weathering method) and fluorescent UV lamps (F) are the most commonly employed artificial radiation sources today. Differences can also be found in the types of optical filters. The most important irradiation parameters dependent on the lamp and the filter are then listed. The next lines show the spectral irradiance at 340 nm, the whole UV irradiance (300 to 400 nm), the UV cut-on, the daily UV radiant exposure, and the time of irradiation. Significant differences can also be found in essential climatic test parameters, such as TOW+U (time of simultaneous action of water and UV), TOW (whole time of wetness), relative humidity during irradiation, and maximum black standard temperature BST_{max}. Finally, the respective exposure cycles are listed, and also exhibit marked differences.

6.2.2 Simulation by means of artificial radiation sources

6.2.2.1 Radiation sources

The search for radiation sources for simulating the action of global radiation in the laboratory mostly focuses on achieving maximum similarity with natural radiation parameters. This means maximum agreement in spectral distribution, such that all global radiation wavelengths which

Table 6.1: Properties of the artificial weathering methods used by [16] and cited as typical methods in the following text

Weathering method	X1a CAM 180 SAE-J 1960	X2a cycle 102:18 ISO 11341	F1a mode 1 (8/4) ISO 4892-3	F2a mode 1 (8/4) ISO 4892-3	F2b mode 1 (4/4) ISO 4892-3
Lamp type	xenon water-cooled	xenon water-cooled	fluorescent UV-A-340	fluorescent UV-B-313	fluorescent UV-B-313
Filter type	quartz/Boro	boro/boro			
UV at 340 nm (W/m²/nm)	0.55	0.35			
UV cut-on (nm)	285	300	300	275	275
UV radiant exposure (MJ/m²/d)	3.7	6.0	3.0		
Time of irradiation (h/d)	16	24	16	16	12
TOW+U (h/d)	2.7	3.6	0	0	0
TOW (h/d)	10.7	3.6	8	8	12
Relative humidity during irradiation (%)	50		5	5	7
BSTmax (°C)	70	63	70	70	60

Exposure cycle	
X1a	40 min dry irradiation, 20 min irradiation with wetness by rain, 60 min dry irradiation, 60 min dark with wetness by rain
X2a	102 min dry irradiation, 18 min irradiation with wetness by rain
F1a, F2a	8 h dry irradiation (70°), 4 h dark with wetness by condensation (50 °C)
F2b	4 h dry irradiation (60°), 4 h dark with wetness by condensation (50 °C)

Table 6.2: Properties of the artificial weathering methods used by [22] and cited as typical methods in the following text

Weathering method	X1a CAM 180 SAE-J 1960	X1b CAM 180 SAE-J 1960	X2b cycle 102:18 ISO 11341	F1b mode 1 (4/4) ISO 4892-3	F2b mode 1 (4/4) ISO 4892-3
Lamp type	xenon water-cooled	xenon water-cooled	xenon air-cooled	fluorescent UV-A-340	fluorescent UV-B-313
Filter type	quartz/boro	boro/boro	Xeno-chrome 300		
UV at 340 nm (W/m²/nm)	0.55		0.35		
UV 300 to 400 nm (W/m²)	64	80	70		
UV cut-on (nm)	285	300	295	300	275
UV radiant exposure (MJ/m²/d)	3.7	4.6	6.0	2.2	
Time of irradiation (h/d)	16	16	24	12	12
TOW+U (h/d)	2.7	2.7	3.6	0	0
TOW (h/d)	10.7	10.7	3.6	12	12
Relative humidity during irradiation (%)	50	50	55	7	7
BSTmax (°C)	70	80	65	60	60

Exposure cycle	
X1a , X1b	40 min dry irradiation, 20 min irradiation with wetness by rain, 60 min dry irradiation, 60 min dark with wetness by rain
X2b	102 min dry irradiation, 18 min irradiation with wetness by rain
F1b , F2b	4 h dry irradiation (60°), 4 h dark with wetness by condensation (50 °C)

exert an influence on coating ageing are considered. Good agreement of the radiation spectra can be expected to yield good agreement in coating ageing. For instance, the presence of radiation of wavelength below 300 nm, which has unnaturally high photon energy, as well as the absence of radiation of wavelength around 320 nm, which greatly influences coating ageing, should be avoided.

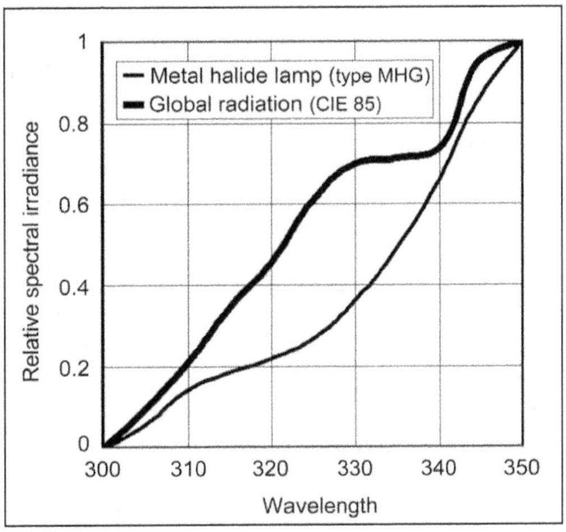

Figure 6.1: Relative spectral irradiance of a typical metal halide lamp (MHG type) in the wavelength range 300 to 350 nm in comparison to that of the global radiation (CIE 85, T.4) (adapted from [71]

According to the laws of radiation physics, the radiation spectrum of black body emitters predominantly depends on their surface temperature. Therefore, arc lights having roughly the sun's surface temperature were chosen first. Only carbon arc and xenon lamps have attained any significance in artificial weathering. The carbon arc lamp was the first type to be used for this purpose, in 1918. Later, it was replaced by the xenon lamp, because the spectrum of xenon arc radiation shows greater similarity to that of global radiation. Nowadays, carbon arc lamps have virtually no significance in artificial weathering. Some 25 years ago, the fluorescent UV lamp appeared on the scene as a competitor for the xenon lamp. However, its radiation is limited to the wavelength range of the UV radiation. Therefore, fluorescent UV lamps are intended to simulate only the global UV that is the most effective wavelength range of the global radiation in terms of coating ageing. Nowadays, only these two types of lamps can be recommended as radiation sources for artificial weathering in coatings testing. Their advantages and disadvantages are compared later in Chapter 6.2.2.3.

A further type of artificial radiation source is the so-called metal halide lamp. It is predominantly used for testing complete technical products, for instance, fully equipped automobiles, railway engines, and passenger trains. Big radiation systems that use metal halide lamps are available. They are typically used to test the effect of the solar heat load of air-conditioning systems. However, metal halide lamps cannot be recommended for testing the ageing behaviour of coatings. Their radiation spectrum differs too much from that of global UV.

Figure 6.1 compares a typical radiation spectrum of a metal halide lamp often used for irradiating fully equipped automobiles with that of global radiation for the wavelength range from 300 to 350 nm. Pronounced deviations can be seen in the range 320 to 330 nm. Especially, spectral irradiance at 320 nm, which is known to have a particular influence on coating ageing, only reaches half the value of global radiation. Note that Figure 6.1 shows a relatively good example. Most metal halide lamps show far less radiation in this critical wavelength range. Furthermore, metal halide lamps are not controllable. Further discussions are devoted only to xenon lamps and fluorescent UV lamps.

6.2.2.2 Simulation by means of artificial radiation sources

As mentioned in Chapter 6.2.2.1 above, only the mostly commonly used artificial radiation sources, xenon lamps and fluorescent UV lamps, will be described in this book. Let us look first at xenon lamps. Besides the photochemically active spectral range of radiation, they emit the entire VIS and NIR ranges. In this connection, the percentage of emitted NIR is higher than that contained in global radiation. In order to avoid overly high thermal stress of the test panels, the lamps, the optical filters, and the other irradiated parts of the weathering device, appropriate filtering and cooling means are indispensable. Air or water is commonly used for cooling. The type of cooling is

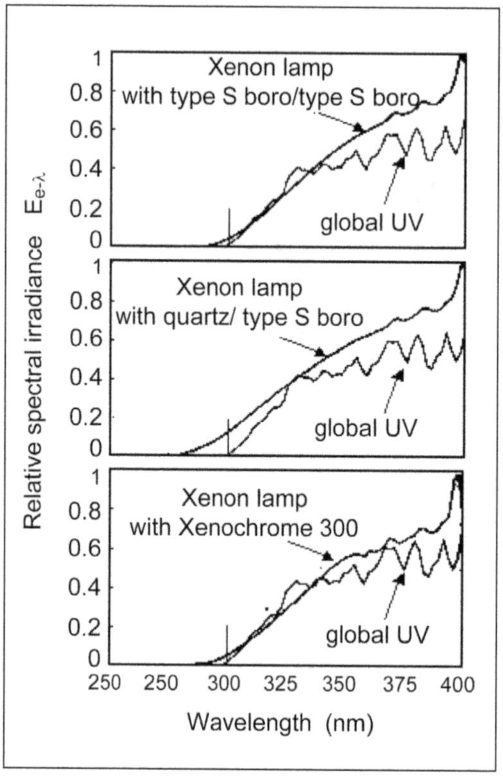

Figure 6.2: Spectral distribution of xenon arc radiation filtered by means of different filter systems in the wavelength range from 250 to 400 nm (adapted from [28])

reported not to have a major influence on spectral distribution, including the cut-on of the UV and therefore, on the weathering results.

Xenon lamps also emit a substantial amount of radiation below the natural cut-on of 300 nm. In order to reliably exclude UV of unnaturally high photon energy, and to decrease the unnaturally high amount of NIR, xenon lamp devices are equipped with optical filter systems. Water-cooled xenon lamps, for instance, are surrounded by two cylindrical optical filters, both preferably made from sufficiently UV-absorbing borate borosilicate glass, type S (boro/boro). Only this type of optical filter reliably avoids irradiation below 300 nm. As can be seen in Tables 6.1 and 6.2, filter combinations containing quartz plus borosilicate glass can still be found in weathering devices used in coatings testing. Quartz/boro filters transmit substantial quantities of UV radiation below 300 nm.

Air-cooled xenon lamp devices are commonly use optical interference filters, for instance, the filter Xenochrome 300. Figure 6.2 shows the spectral distribution in the wavelength range below 400 nm for typical water-cooled and air-cooled xenon lamps as achieved by means of different optical filters.

As may be seen, the radiation emitted from the xenon lamp does not contain any substantial amount below 300 nm after passing through the filter systems boro/boro and Xenochrome 300. In both cases, the test panels are not exposed to radiation of higher photon energy than that of the global radiation. Xenon lamp weathering devices fitted with these types of filter also fulfil the recommendations of ISO 4892-2-2006 concerning admissible deviation from the global radiation, as described in Table 6.3.

However, the substantial amount of radiation below 300 nm can no longer be regarded as norm-compliant when the weathering device is fitted with filter systems of the quartz/boro type, which are still often found in older xenon lamp devices. Therefore, quartz/boro filters should generally be replaced by boro/boro filters as soon as possible. For special irradiation purposes, for instance, to simulate radiation stress behind window glass, filters are available that shift the radiation's cut-on to a correspondingly higher wavelength.

Table 6.3: Standardized relative spectral irradiance for artificial weathering by means of xenon lamps as per ISO 4892-2 method A

Spectral range Wavelength λ, nm	Minimum %	Reference: CIE 85, table 4 (1989) %	Maximum %
$\lambda < 290$		0	0.15
$290 \leq \lambda \leq 320$	2.6	5.4	7.9
$320 < \lambda \leq 360$	28.2	38.2	39.8
$360 < \lambda \leq 400$	54.2	56.4	67.5

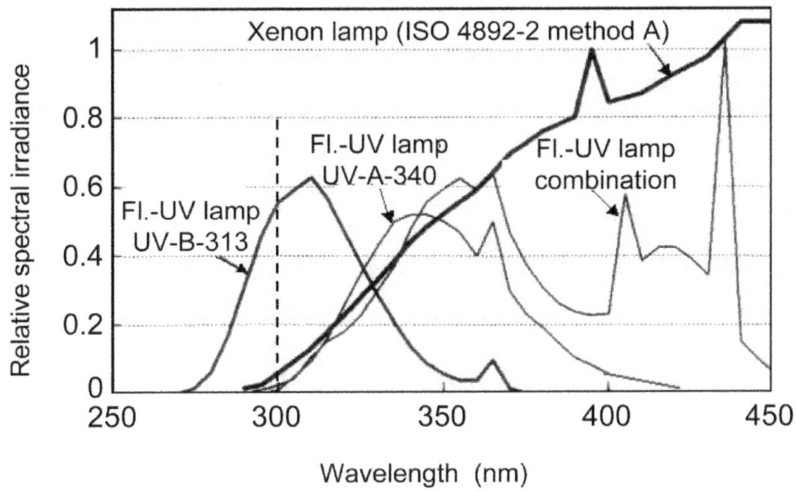

Figure 6.3: Spectral distribution of the radiation generated by different types of fluorescent UV lamps working alone or in combination (as per ISO 4892-3) in comparison to that of filtered xenon arc radiation (as per ISO 4892-2 method A) for the wavelength range 250 to 450 nm (adapted from [42])

Now, let us deal with the fluorescent UV lamp, which is the other most commonly employed radiation source. It was introduced into artificial weathering about 25 years ago. It works on the lines of the fluorescent lamps used for illumination. Appropriate types were developed to restrict as far as possible the emitted radiation to the UV wavelength range. Figure 6.3 shows radiation spectra of the two most commonly employed types of fluorescent UV lamps (UV-B-313 and UV-A-340) and the common combination of four different fluorescent UV lamps relative to that of filtered xenon arc radiation.

As may be seen, the UV-A-340 lamp and the lamp combination show good agreement with the global radiation in terms of the cut-on at 300 nm. The recommendations of ISO 4892-3 concerning permissible deviation from global radiation are shown in Table 6.4. Fluorescent UV lamp weathering devices that employ UV-A-340 lamps and the aforementioned lamp combination may be considered norm-compliant.

As may also be seen in Figure 6.3 the curves of the spectral distribution of the radiation emitted by the UV-A-340 lamp as well as the fluorescent UV lamp combination show quite good agreement with that of the global UV, but only up to 350 nm. Single fluorescent UV lamps are always characterized by their bell-type radiation spectrum. Only the combination of fluorescent UV lamps with a radiation maximum at different wavelengths permits a better approximation to global UV.

Table 6.4: Recommended relative spectral irradiance for artificial weathering by means of UV-A-340 fluorescent UV lamp, or a combination of different fluorescent UV lamps as per method A of ISO 4892-3 in comparison to that of global radiation (CIE85, Table 4)

Spectral range	**Lamp U-VA-340**		**Reference: CIE 85, table 4 (1989)**	**Combination of lamps**	
Wavelength λ, nm	minimum %	maximum %	%	minimum %	maximum %
λ < 290		0.01	0		0
290 ≤ λ ≤ 320	5.9	9.3	5.4	4	7
320 < λ ≤ 360	60.9	65.5	38.2	48	56
360 < λ ≤ 400	26.5	32.8	56.4	38	46

Figure 6.3 also shows the curve for the fluorescent UV-B-313 lamp. As can be seen, this lamp emits a substantial amount of radiation below 300 nm. Method C of ISO 4892-3-2006 also describes a procedure that involves working only with this type of lamp. In this case, the irradiation stress is exceeded by the enhanced photon energy. Although this method fulfils the criteria for an accelerated testing test method as described in Chapter 2.5, care is required in the interpretation of its test results. As will be discussed in more detail later in Chapter 6.2.2.4, irradiation with unnaturally exceeded photon energy can lead to false predictions for weatherability! **If possible, avoid using a UV-B-313 lamp.**

6.2.2.3 Fluorescent UV lamp versus xenon lamp

The question as to "Which technique is the best for my task?" always arises when a testing laboratory is established, extended, or modernized. When ordering a new weathering device, the test engineer has mostly to decide between a xenon lamp and a fluorescent UV lamp.

As mentioned before in Chapter 6.2.2.2, xenon arc radiation contains a high amount of infrared (NIR) which generates heat that has to be removed from the weathering device, mostly in the form of useless energy. For this purpose, effective cooling systems are indispensable. Furthermore, xenon lamps emit visible light, whose contribution to coating photooxidation, and therefore the need for its presence in artificial weathering, is debatable. The latest experience in coatings testing suggests that only UV radiation is capable of degrading common coating resins. Only printing inks and coatings certain organic pigments which are known to suffer photochemical ageing upon irradiation with visible light. Photons always need a minimum of energy before they can break certain chemical bonds. Consequently, chemical bonds that are broken by photons of visible light will also be broken by photons of UV, which are more energetic.

Experience in the use of the reference coating ORWET, containing the light-sensitive organic pigment Ciba "Irgalith" Orange F2G, as described in Chapter 4.4, corroborates this assertion. ORWET uses the degree of pigment destruction as the ageing criterion for measuring such a change in colour. Its activation spectra, determined for irradiation with the fluorescent UV-A-340 and UV-B-313 lamps are presented in Figure 6.4. The spectra demonstrate, first, that UV radiation alone is able to destroy the pigment and, second, that the lamp emitting radiation of the higher photon energy (UV-B-313) effects a greater change in colour for identical radiant exposure. This proves that the additional presence of VIS, as in the case of xenon arc radiation, is not necessary for photochemical ageing of the pigment. Therefore, **photochemical degradation of coating components by visible light is not an argument against the fluorescent UV lamp,** especially if the photochemical degradation, irrespective of whether it is caused by VIS or UV, leads to the same ageing effect (in this case a change in colour).

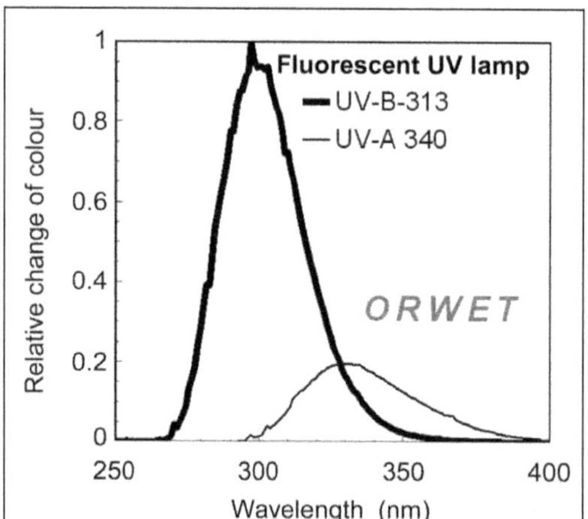

Figure 6.4: Activation spectra of the ORWET reference coating resulting from irradiation with different types of fluorescent UV lamp (UV-B-313 and UV-A-340) in which colour change serves as ageing criterion (simplified presentation without spectral lines, adapted from [72])

However, xenon lamps have the advantage that greater photochemically active irradiance is available when compared with fluorescent UV lamps, on the basis of identical spectral UV irradiance. This may be clearly seen in Figure 6.5, where ORWET's activation spectrum, determined by irradiation with the UV-A-340 fluorescent lamp (as per ISO 4892-3), is compared with that determined by irradiation with the xenon lamp (as per ISO 4892-2 method A). The area beneath the curve for the xenon lamp in the range 300 to 500 nm is almost twice as large as that beneath the curve for the fluorescent UV lamp. If both lamps are controlled on the basis of identical spectral irradiance at 340 nm, as is commonly practised, ORWET can be expected to change its colour twice as much upon irradiation with xenon lamps than upon irradiation with fluorescent UV lamps.

In most cases, irradiation with UV is sufficient to determine the ageing behaviour of modern coatings. Therefore, the question arises as to which is more economical. Since no energy is wasted in generating VIS and NIR, fluorescent UV lamps have a clear advantage in terms of economics. For the same UV irradiance, fluorescent UV lamp devices consume only about a sixth of the cost of electrical energy and only about a third of the costs of regular replacement of lamps in comparison to xenon lamp devices.

When deciding between a xenon lamp and a fluorescent UV lamp, remember that a **xenon lamp device is at a clear disadvantage in terms of running costs if irradiance**, climatic performance, and test panel capacity are identical. Especially, a general desire to save electrical energy and other running costs raises the question: "Can xenon lamps be replaced by fluorescent UV lamps without any restrictions on the prediction of coatings' service lives?" There is no unequivocal answer at present. Although fluorescent UV lamps are very good at simulating the most ageing-effective wavelength range of 300 to 350 nm, they cannot adequately simulate the wavelength range 350 to 400 nm in terms of global UV. Due to their bell-type radiation spectrum, fluorescent UV lamps' irradiance diminishes markedly with increasing wavelength above 350 nm, in contrast to global UV. This may be clearly seen in Table 6.5. It compares the ratio of spectral irradiances at 370 nm and 320 nm for xenon arc radiation, fluorescent UV lamp radiation, and global radiation. This ratio is roughly equivalent to the ratio of UV irradiance above 350 nm to that below 350 nm.

The UV-A-340 fluorescent lamp has a ratio of about 1; the xenon lamp, of about 3; and the global UV, of about 2.4. Thus, the different radiation sources clearly differ in their contribution to the wavelength range above 350 nm. In comparison to global UV, the fluorescent UV lamp emits less and the xenon lamp more radiation above 350 nm. Although radiation below 350 nm is generally considered the most ageing-effective wavelength range, UV radiation above 350 nm is also capable of breaking certain chemical bonds, as was seen in Figure 3.3. Inexplicable differences in ageing behaviour that occur after irradiation with different radiation sources may be caused by unequal spectral distribution inside the UV wavelength range 300 to 400 nm.

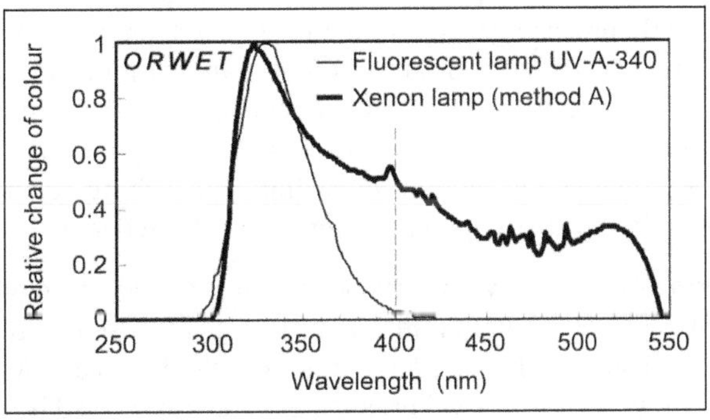

Figure 6.5: Activation spectra of the ORWET reference coating resulting from irradiation with the UV-A-340 fluorescent lamp as per ISO 4892-3, in comparison to that resulting from irradiation with the xenon lamp as per ISO 4892-2 method A, where colour change serves as ageing criterion (simplified presentation without spectral lines, adapted from [72])

Table 6.5: Ratio of the irradiance acting on the coating surface in the narrow wavelength range of 370 ± 0.5 nm ($E_{e\text{-}370}$) to that of 320 ± 0.5 nm ($E_{e\text{-}320}$) as a function of the source of radiation used for weathering

Quotient Ee-370/Ee-320		
Global radiation	CIE 85, table 4	2.4
Filtered xenon arc radiation	ISO 4892-2-method A	3
Radiation of the fluorescent UV lamp type U-VA-340	ISO 4892-3-type 1	1

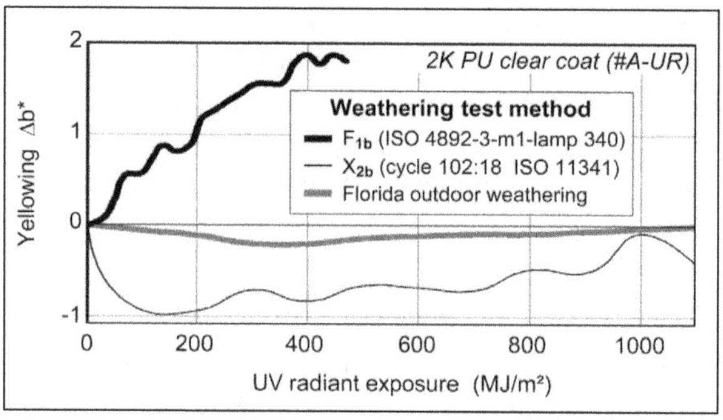

Figure 6.6: Yellowing (Δb^*) of a well stabilized 2-pack PUR clearcoat applied to a white basecoat as a function of the received UV radiant exposure and the weathering test method employed; see Table 6.2 for more details (adapted from [22])

A corresponding example found in the literature [22] is presented in Figure 6.6. It shows the extent of the different action of global radiation, xenon arc radiation and fluorescent UV lamp radiation on the yellowing of a basecoat/clearcoat system for automotive OEM, consisting of a well stabilized 2-pack PUR clearcoat applied to a white basecoat.

The weathering test method "F_{1b}", which utilizes fluorescent UV lamps, produces yellowing increases continuously with increasing radiant exposure. In contrast, filtered xenon arc radiation, used in the "X_{2b}" weathering test method, causes bleaching of the coating system, and undefined falling and rising with increasing radiant exposure. Bleaching is characterized by negative values on the b* yellow/blue coordinate of the CIELab colour system. Using the sun as natural radiation source failed to lead to considerable yellowing or bleaching, as the curve for conventional outdoor weathering in southern Florida shows.

In Figure 6.6 and Table 6.5, the ageing behaviour produced by weathering tests involving different radiation sources agrees very well with the different ratios of UV radiation above and below 350 nm. This relationship may be the reason for the different weathering results. This phenomenon might be explained as follows. It is known that the photochemical degradation of polymers mostly proceeds via a number of steps which generate more or less stable interim reaction products. Such interim products often manifest themselves in a change of colour. As already shown for the example of PVC in Chapter 3.2.3, weathering can often lead to yellowing or bleaching, depending on the weathering conditions. In the example of PVC, differences in water stress were responsible for whether yellowing or bleaching occurred. In the case of our clearcoat, different irradiation stress (photon energy) may have decided whether yellowing occurred or not.

Probably, only UV photons below 350 nm had sufficient energy to break chemical bonds, leading to a yellow-coloured interim product. Let us assume that this yellow species was less stable than the original polymer, and therefore could be broken by all UV photons irrespective of their energy. In other words, the yellow species could be cleaved by UV radiation below as well as over 350 nm. Suppose that cleaving of the yellow species led to colourless further reaction products. It is also known that yellowing can be detected by standard instruments only after a certain concentration of yellow-coloured molecules has been reached. So, in our example, bleaching upon irradiation with the xenon lamp might be caused by the fact that the molecules of the yellow species were cleaved faster than they were generated due to the presence of relatively high irradiance above 350 nm. In that event, the concentration necessary to make the yellow species visible might not have been reached. In contrast, fluorescent UV lamps adjusted to the same irradiance in the range 300 to 400 nm provide considerably less UV radiation above 350 nm. In this case, more molecules of the yellow species might be generated than degraded at the same time, and the concentration necessary for rendering the yellowing visible might be achieved. **In cases where the results of artificial weathering carried out in fluorescent UV lamp devices failed to correspond with ageing behaviour in normal use, xenon lamp devices have justified their retention.**

A further argument often adduced for or against the xenon lamp are degradation processes that lead to irreversible changes in spectral transmittance during weathering of transparent polymers. The spectral transmittance of common clearcoats is known to increase in the wavelength range below 350 nm during weathering, due to the photochemical degradation of UV absorbers. Figure 3.10

shows for the example of clearcoat #01 that the spectral transmittance at 310 nm increased about 20 % after one year of outdoor weathering in Florida. Figure 6.7 shows a further example of a different clearcoat. It depicts the curves of the spectral transmittance of a non-weathered specimen and a specimen weathered outdoors in southern Florida. As in the aforementioned example, more energy-rich radiation penetrates deeper into the coating layer with increasing radiant exposure. This behaviour will be particularly important if the ability of clearcoats to protect the basecoats must also be tested. Clearcoats with this kind of ageing behaviour should preferably be tested in weathering devices that employ fluorescent UV lamps.

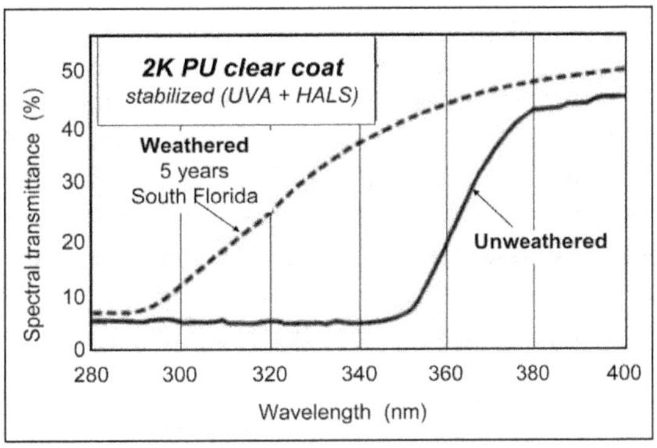

Figure 6.7: The influence of 5-year conventional outdoor weathering in southern Florida on the spectral transmittance of a stabilized 2-pack PUR clearcoat (adapted from [16])

However, there are also transparent polymers that exhibit reverse ageing behaviour. With these polymers, weathering leads to interim ageing products in the outermost micrometers of their surface that are very stable to further photochemical ageing and act like an optical filter, especially for shortwave UV. Such layers prevent further penetration of UV into the polymer and virtually terminate finish further photochemical ageing. Polycarbonate is a typical example of this kind of ageing behaviour. Figure 6.8 shows the spectral transmittance of a transparent non-stabilized polycarbonate panel, measured in the wavelength range 280 to 440 nm, before and after weathering. As can be seen, 25 % of the penetrating radiation at 320 nm passes through the plate in its unweathered state. However, after weathering, radiation of 320 nm wavelength could no longer be measured behind the panel.

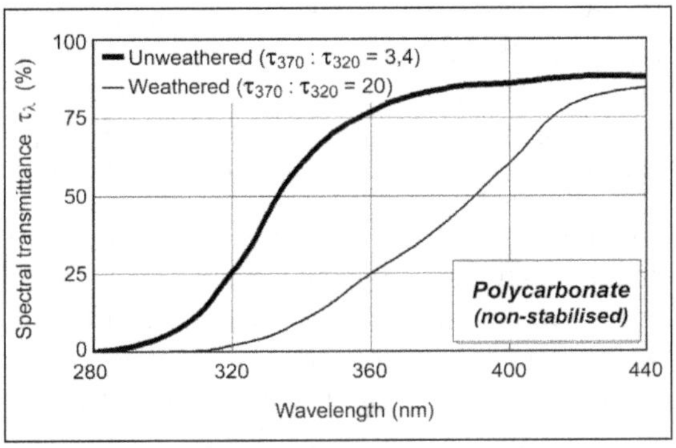

Figure 6.8: Weathering-induced change in the spectral transmittance and the ratio of transmittance at 370 nm to that at 320 nm for non-stabilised, transparent PC (from [72])

The ratio of the radiation, passing through the plate at 370 nm to the radiation passing through it at 320 nm, changes dramatically due to weathering. On account of this ageing behaviour, further ageing would virtually cease if fluorescent UV lamps were used for weathering. The most effective portion of UV emitted by fluorescent UV lamps below 350 nm is strongly decreased by absorption in the outermost surface layer. That portion of UV remaining above 350 nm, which undergoes less absorption, is only small, due to the bell-type radiation spectrum. The xenon lamp might be a better radiation source for clearcoats exhibiting this kind of ageing behaviour, because a higher amount of UV radiation in the wavelength range above 350 nm is available for weathering.

The curve for the combination of different types of fluorescent UV lamps, also presented in Figure 6.3, shows increased UV irradiance above 350 nm, but does not attain the level of the global radiation. Although offering advantages over the UV-A-340 lamp, the lamp combination cannot be used in all types of weathering devices that employ fluorescent UV lamps, because of design reasons.

Besides the economical advantage, fluorescent UV lamps have a further important advantage. Since the emitted radiation is limited to the UV range, which constitutes only about 6 % of xenon arc radiation, fluorescent UV lamps generate commensurately less thermal stress than xenon lamps. Therefore, the increase in the test panel temperature caused at the start of irradiation can be neglected. In other words, the surface temperature during weathering by fluorescent UV lamp devices is almost independent of the reflectance of the test panels. The coating surface micro climate, which is known to have a marked influence on coating ageing, can be kept at the same level when test panels of different colour are weathered. **Proponents of xenon lamps often advance the argument that weathering in xenon lamp devices best simulates the natural weathering stress of differently coloured test panels.** However, this argument does not hold for all of the most important weather situations occurring in normal use.

Figure 6.9 illustrates the case of four differently coloured test panels and shows how well the two artificial radiation sources simulate the thermal stress in the two most important weather scenarios occurring in Miami, Florida. The two chosen weather scenarios, shown in the two rows of pictures, represent the worst case weather scenarios d and e, as per Figure 3.46. The first picture in the top row shows how the colour of the test panels influences their surface temperature in weather scenario d. Surface temperatures up to 70 °C for black test panels can be observed outdoors on the black box in southern Florida under clear skies at noon in June. Clear skies means that irradiation predominantly by direct solar radiation is responsible for the differences in surface temperature. The next two pictures in the top row show the colour-dependent differentiation for irradiation with a xenon lamp device and with a fluorescent UV lamp device. In both weathering devices, the test parameters were adjusted to allow the black-coloured panels to reach 70 °C.

As can be seen in this row, the xenon lamp is very good at simulating the coating surface micro climate which occurs under clear skies in southern Florida due to the extreme of direct solar radiation. The influence of colour on a coating surface micro climate and therefore on coating ageing in a weather scenario characterised by a high amount of direct solar radiation found under clear skies cannot be reasonably simulated by weathering in a fluorescent UV lamp device.

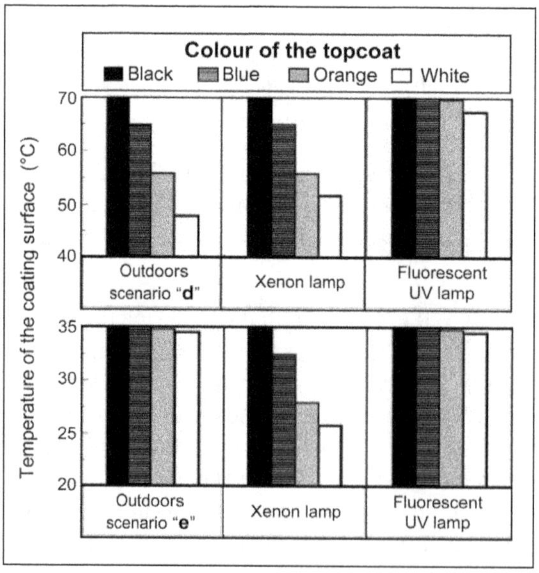

Figure 6.9: Surface temperatures of differently coloured specimens of automotive coatings during irradiation with different artificial radiation sources in comparison to those of two typical weather scenarios frequently occurring during outdoor weathering on the black box in Miami, Florida at noon on the same June day (d: clear sky; e: sun temporarily covered by a cloud)

However, as mentioned before, **clear skies feature in just one of the most important weather scenarios and need not be the worst case weather scenario for the coatings to be tested**. Especially during the cloudier summer months in southern Florida, weather scenario e occurs there nearly as frequently as weather scenario d. Weather scenario e occurs precisely when a cloud covers the sun and prevents direct solar radiation from reaching the test panel surface. Due to the interruption in the direct irradiation, especially of the direct portions of VIS and NIR, the temperature of the test panels immediately drops to virtually the ambient temperature.

The relative humidity of the coating surface micro climate is almost identical with that of the open air. In this situation, only the diffuse portion of the solar radiation coming from the full half space of the sky reaches the test panels. Since scattering, as the reason for the diffuse radiation, increases with falling wavelength, the diffuse portion can constitute as much as 50 % of the whole global UV, as already discussed in Chapter 3.2.1.4. In other words, in spite of the clouds, the coatings to be tested are irradiated with much the same radiation in the most ageing-effective wavelength range. As described in Chapter 3.2.5, this weather situation could also be construed as the possible worst case weather scenario for coatings. Since a degree of cloud cover of 50 % on average can be assumed during the summer season in southern Florida, weather scenarios e and d are expected to occur with about the same frequency.

The bottom row on Figure 6.9 simulates the natural stress in weather scenario "e" produced by the two artificial radiation sources. The type of presentation is identical with that for weather scenario d. The first picture on the bottom row shows that the diffuse portion of the global radiation does not generate any considerable colour-related differences in the test panel temperature, which is almost identical with that of the ambient air, namely 35 °C (also at noon in June). In both weathering devices, test parameters were adjusted to let the temperature reach 35 °C on the black-coloured test panels. Comparison of the three pictures in the bottom row shows that **only the fluorescent UV lamp is able to correctly simulate a worst case weather scenario characterised by UV irradiation at very high relative humidity and subtropical ambient temperature.** A coating surface micro climate that is equal for all panels and independent of their colour cannot be realized sensitively by xenon lamps because of their high thermal load.

As discussed before, both of the aforementioned weather scenarios occur to about the same extent in normal use. Therefore, in comparison to normal use, dark-coloured coatings are overly stressed in xenon lamp devices and under-stressed in fluorescent UV lamp devices. Light-coloured coatings are stressed in the opposite way.

Furthermore, xenon lamp devices are also disadvantaged by the strongly limited climatic working and controlling range, due to the high thermal load of the xenon lamps. The limitation on the possibility of controlling the coating surface micro climate can be seen in Figure 6.10. It shows the climate working and controlling range for three types of modern xenon lamp devices Ci3000, Ci4000 und Ci5000, which differ mostly in their testing capacity.

The area enclosed by the corresponding curves is commonly referred to as the climate working and controlling range. As may be seen, the surface temperature of dark-coloured coatings, roughly expressed by BPT/BST, is limited on the downside to about 60 °C in the Ci5000, and to 45 °C in the Ci3000 and the Ci4000, due to the high thermal load of the xenon lamps. Weathering at deeper surface temperature is not possible. The relative humidity of the coating surface micro climate can be roughly calculated for dark-coloured coatings from the values of the BPT/BST and the relative humidity of the chamber atmosphere. If a usual difference of 20 °C is assumed between the chamber temperature and the coating surface temperature, it is not possible to work with a relative humidity of the coating surface micro climate that is higher than 30 % in all three weathering devices. Weather scenarios characterized by UV irradiation at almost 100 % relative humidity and temperatures below 40 °C, which are also typical for southern Florida, cannot be adequately simulated in xenon lamp devices. Figure 6.11 (page 126) shows the climatic working and controlling range for a GUT200 fluo-

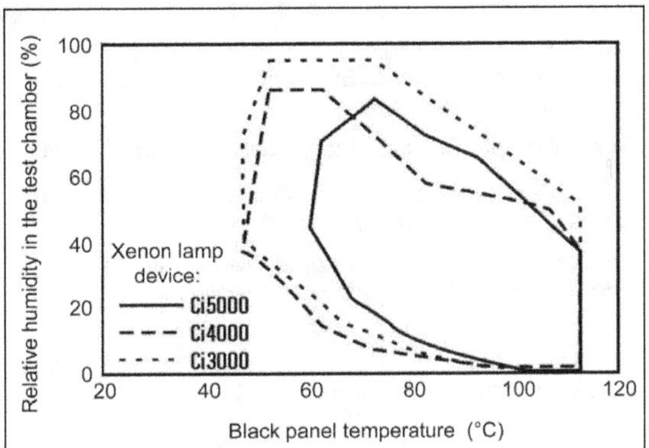

Figure 6.10: Climatic working and controlling range of three representative xenon lamp devices (Ci3000, Ci4000 and Ci5000) with regard to BPT and relative humidity in the chamber (from [74])

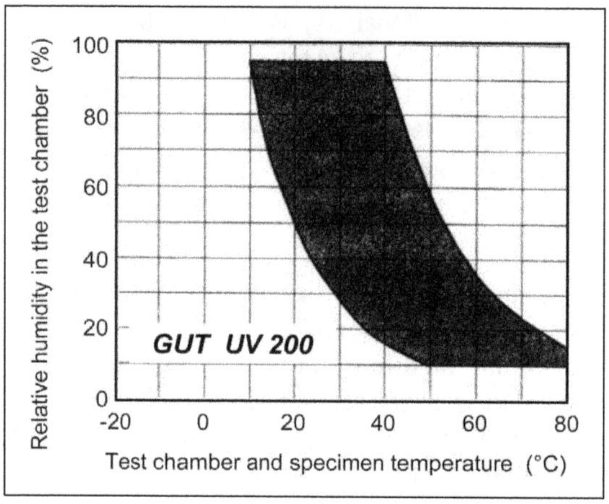

Figure 6.11: Climatic working and controlling range of a GUT200 fluorescent UV lamp device in terms of temperature and relative humidity of the chamber atmosphere, which are almost identical with those of the test panel surface (herein represented by the black area (from [75]))

rescent UV lamp device. As may be readily seen, the aforementioned weather scenario can be achieved in weathering devices by means of fluorescent UV lamps, without any problems or restrictions.

Due to the low irradiation-induced thermal stress, fluorescent UV lamp devices are ideal for separate studies of the influences of individual weathering stresses on coating ageing, such as temperature, relative humidity, acid atmospheric precipitation, and biological attack. For this reason, fluorescent UV lamp devices are preferred in research, for instance, in the development of new weathering test methods or new coatings.

In order to show how precisely the spectral distribution of global radiation must be simulated, the spectral sensitivity of the coatings to be tested should be determined beforehand. If VIS or UV >350 nm do not contribute substantially to coating ageing or lead to the same ageing effects as UV <350 nm, fluorescent UV lamp devices should be preferred; this also applies to the monitoring of current series production.

In summary, Table 6.6 compares the two artificial radiation sources with regard to the most essential weathering aspects. This Table is intended to help the staff involved in coatings testing to make the right decisions concerning that weathering test method and radiation source which best suits their specific task.

Table 6.6: Comparison of xenon lamp and fluorescent UV lamp

Aspect	Xenon lamp	Fluoreszcent- UV lamp (used as reference)
Price of the weathering device (basis: identical maximum UV-irradiance, identical climatic performance, identical total exposure area)	about the same	
Running costs (relative units): a) electric energy b) regular change of lamps and filters	 a) approximately: 6 b) approximately: 3	 a) 1 b) 1
Simulation of the global UV: a) spectral range 320 to 350 nm b) spectral range 350 to 400 nm	 a) very good b) very good	 a) very good b) insufficient
Simulation of the VIS of the global radiation	very good	not possible
Influence of lamp ageing on the spectral distribution of the emitted radiation	considerable	small
Simulation of the colour-related thermal stress caused by the direct portion of the global radiation	very good	not possible
Simulation of the coating surface micro climate	only for weather situations with direct radiation	convenient for all weather situations, especially such with diffuse radiation
Limits of the artificial coating surface micro climate on dark-coloured coatings	minimum temperature: 45 °C maximum rel. humidity: 30 %	minimum temperature: - 10 °C maximum relative humidity: 95 % (till 40 °C)
Maximum UV irradiance (300 to 400 nm)	approximately 180 W/m²	approximately 70 W/m²

6.2.2.4 *Acceleration by enhanced photon energy*

Frequently, the test engineer is often confronted with the opinion: "Anything photons of global radiation can do, photons of higher energy (smaller wavelength) can do much better". Proceeding from the experience that the ageing rate of polymers generally increases with decreasing wavelength, there have been attempts to use radiation of higher photon energy to accelerate artificial weathering of coatings. As already mentioned, in Chapter 6.2.2.1, coatings to be tested are extensively stressed by radiation below the natural cut-on of 300 nm if weathered in certain weathering devices. As may be seen in Figure 6.2, xenon lamp devices equipped with quartz/boro filters employ a considerable amount of UV below 300 nm, whose influence on coating ageing cannot be neglected. The situation will be more dramatic in the case of fluorescent UV lamp devices which are equipped with UV-B-313 lamps. More than 10 % of the radiation impinging on the test panels has a wavelength below 300 nm. Automotive coatings are reported to attain the ageing criterion crack formation up to 50-times faster when exposed to this fluorescent lamp than to a UV-A-340 lamp. Shortening the time outlay to such a high extent, whether for the purpose of developing of new coatings or for monitoring series production, has always been the wish of test engineers.

Figure 6.12 shows how the presence of radiation of wavelength below 300 nm during weathering influences the coatings' ageing behaviour, for the example of an unstabilised 2-pack PUR clearcoat on a white basecoat exposed to weathering test methods employing radiation both with and without wavelength portions below 300 nm. For better comparison, identical weathering cycles were used for each kind of lamp to achieve identical stress caused by the other weathering stress factors, such as heat and water. Therefore, different ageing behaviour revealed by the use of different types of xenon lamps or different types of fluorescent UV lamps might also be related to the differences in the position of their cut-on inside the radiation spectrum.

Yellowing and loss of gloss, serving as ageing criteria, were tracked for up to 2000 hours of exposure. As shown in the lower diagram of Figure 6.12, only those test methods employing the lower cut-on (X_{1a} at 285 nm and F_{2b} at 275 nm) showed a considerable loss of gloss. The presence of radiation below 300 nm led to much higher loss of gloss (about 40 % for F_{2b} and 80 % for X_{1a}) relative to the tests methods that only employed radiation >300 nm (X_{1b} and F_{1b}) and showed no loss of gloss at all. With regard to yellowing, the influence of lower cut-on of UV-B-313 lamp (F_{2b}) manifested itself in about three-times as much yellowing relative to the UV-A-340 lamp used for the F_{1b} method (see upper diagram in Figure 6.12).

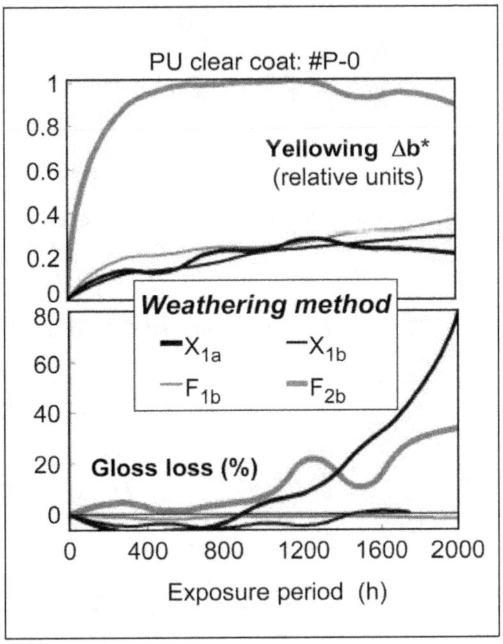

Figure 6.12: Yellowing and loss of gloss by an unstabilised polyurethane clearcoat applied to white basecoat during artificial weathering by radiation sources with (X_{1a} and F_{2b}) as well as without (X_{1b} and F_{1b}) UV radiation below 300 nm. The four weathering test methods employed are explained in detail in Table 6.2 (adapted from [22])

Figure 6.13 shows the influence of photon energy for a further example. Series of four differently stabilized 2-pack PUR clearcoats were weathered in parallel in weathering devices employing different types of fluorescent UV lamps (UV-A-340 and UV-B-313). The exposure period needed for yellowing of $\Delta b^* = 2$ served as the ageing criterion. Comparison of the two diagrams in Figure 6.13 reveals that test method F_{2b}, employing lamp UV-B-313 which

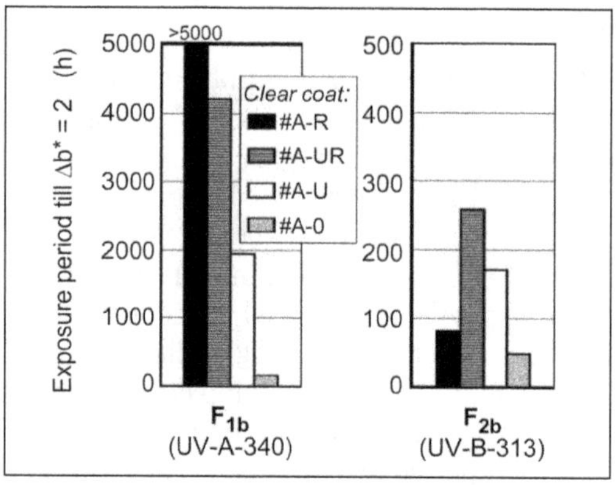

Figure 6.13: Exposure period up to yellowing of Δb^* = 2 for four differently stabilized samples of a 2-pack PUR clearcoat applied to white basecoat subjected to fluorescent UV lamps with (F_{2b}) and without (F_{1b}) radiation below 300 nm. The weathering test methods employed are explained in detail in Table 6.2 (adapted from [22])

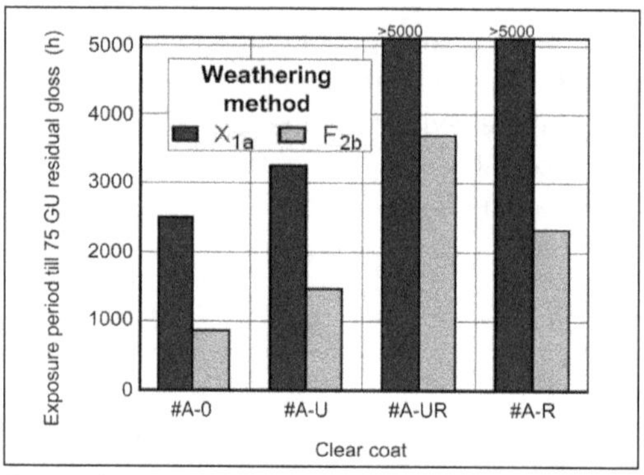

Figure 6.14: Exposure period up to a residual gloss of 75 units for four differently stabilized samples of a 2-pack PUR clearcoat applied to a white basecoat and exposed to xenon lamps (X_{1a}) as well as fluorescent UV lamps (F_{2b}), all emitting radiation below 300 nm. The weathering test methods employed are explained in detail in Table 6.2 (adapted from [22])

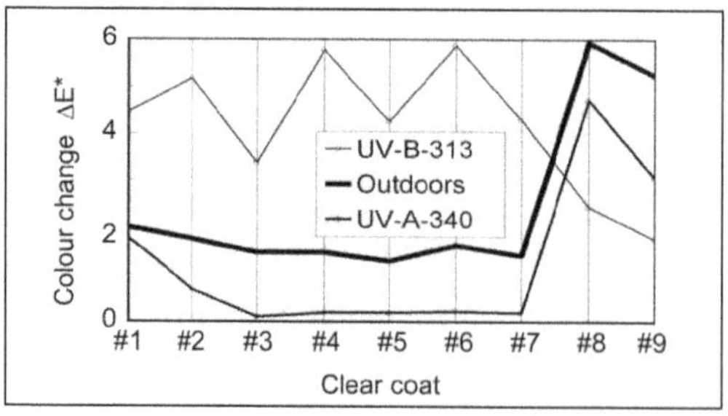

Figure 6.15: Weathering-induced change in colour ΔE^* in weathering devices employing fluorescent UV lamps emitting radiation below 300 nm (UV-B-313) or not (UV-A-340), in comparison to that of conventional outdoor weathering in southern Florida for the example of a series of 9 coatings of different weatherability (adapted from [76]).

emitted the photons of higher energy, led to 50-times faster coating ageing than method F_{1b}, whose UV-A-340 lamp emitted photons of lower energy.

Figure 6.14 uses the example of the same four clearcoats to illustrate the different effects of the tests methods X_{1a} and F_{2b}, both employing radiation having wavelength portions below 300 nm. The exposure period necessary for reducing the residual gloss down to 75 gloss units served as the ageing criterion. The very good ranking correlation between the weathering results of two test methods employing unnaturally high photon energy was surprising. Since the two clearcoats that were tested differed only in their stabilization, clearly the effect of the radiation below 300 nm predominated.

However, radiation below 300 nm could set ageing processes in motion that would never occur in normal use. For this reason, artificial weathering of coatings of different weatherability could yield rankings which differ from normal use. This assertion is corroborated by a typical example published by [76] and presented in Figure 6.15. It shows the results of conventional outdoor weathering performed in southern Florida as well as the results of artificial weathering performed in fluorescent UV lamp devices employing different types of lamp. Nine different polyester-based clearcoats (#01 to #09) were involved in these studies. Change in colour served as the ageing criterion.

A look at the curves shows that only the results of artificial weathering by means of the UV-A-340 lamp correlate with the results of outdoor weathering. In contrast, weathering with a UV-B-313 lamp fails to correlate with the two other tests. The clearcoats that proved to be the best in outdoor weathering proved the worst in artificial weathering with a UV-B-313 lamp. Similar conclusions were drawn in studies of two differently based powder coatings, as presented in Figure 6.16. Each of the two weathering test methods revealed a different coating to be the best one.

Furthermore, UV radiation of unnaturally high photon energy also proved to accelerate the degradation of the commonly employed light stabilisers. Stabilisers of the HALS type are characterized by their property to resist global radiation stress without suffering any damage. As they were developed to protect the coating only against the real weathering stresses, the property whereby HALS molecules are regenerated is lost when they are irradiated with photons of unnaturally high energy. Figure 6.17 shows for the example of a commonly stabilized 2-pack PUR clearcoat the change in content of HALS, expressed by the concentration of nitroxyl free-radicals, over the course of the two artificial weathering test methods X_{1a} and F_{2a} differing in the cut-on of the used radiation. The curve for the X_{1a} method proved that no HALS molecules were degraded by radiation down to 285 nm. In contrast, radiation with a cut-on just 10 °C lower (275 nm) led to considerable loss of HALS. As shown by the curve for the F_{2a} method, the clearcoat lost about 50 % of its initial HALS during 1000 hours of exposure, due to unnaturally enhanced photon energy.

Furthermore, UV absorbers are also known to degrade faster as the photon energy increases. As a result, the ratio of the rate of polymer degradation to the rate of loss of UV absorbers in normal use, which serves as basis for selecting the most suitable UV absorber and its content, will be not correctly simulated by artificial weathering if radiation of unnaturally high photon energy is used. Figure 6.18 shows the influence of the photon energy on the exposure period necessary for achieving 50 % loss of UV absorber for the example of two differently stabilized samples of a 2-pack PUR clearcoat exposed to fluorescent UV lamps with different cut-on. As shown in this figure, both clearcoats experienced about 40 % faster destruction of UV absorber when exposed to the UV-A-313 lamp emitting radiation below 300 nm as well.

The discussions in this Chapter of the book make it clear that **weathering test methods employing lamps that also emit radiation of unnaturally high photon energy (wavelength below 300 nm) cannot be used for differentiating coatings in terms of their ageing behaviour in normal use**. Otherwise, it could not be ruled out that newly created coatings that would behave excellently in nor-

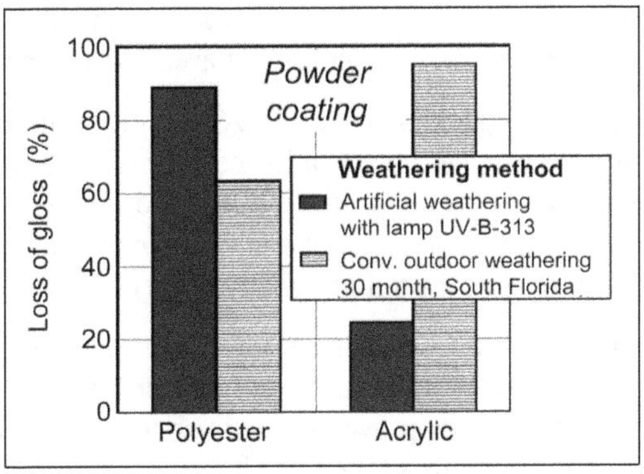

Figure 6.16: Loss of gloss caused by 400 hours of artificial weathering in a device employing a UV-B-313 fluorescent lamp in comparison with that caused by 30 months of conventional outdoor weathering in southern Florida for the example of a series of two powder coatings based on different resins, adapted from [28]

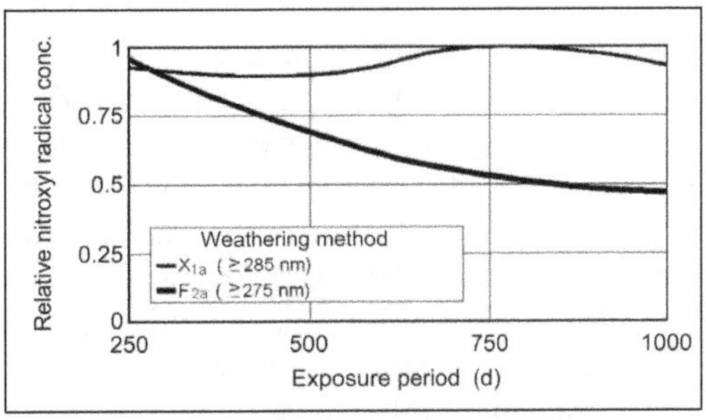

Figure 6.17: Effect of UV radiation of different photon energy on the content of HALS (expressed here as the concentration of nitroxyl free-radicals) for the example of an acrylate/isocyanate 2-pack-polyurethane clearcoat artificially weathered by test methods differing in the cut-on of the used radiation. The test methods employed are explained in detail in Table 6.1, adapted from [16]

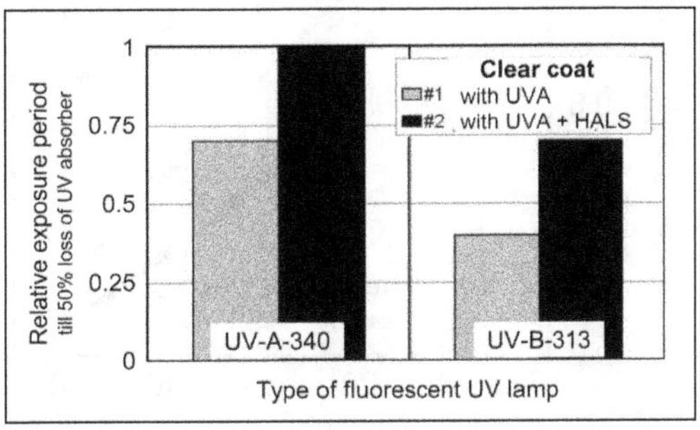

Figure 6.18: The influence of photon energy on the loss of UV absorbers for the example of two differently stabilized samples (with UVA or with UVA+HALS) of a 2-pack PUR clearcoat artificially weathered in fluorescent UV lamp devices employing UV-A-340 and UV-B-313 lamps that emitted radiation of different cut-on, adapted from [16]

mal use would be prevented from reaching series production because the artificial weathering test methods were unsuitable for correctly simulating the real weathering stress.

If possible, avoid using a UV-B-313 lamp at all. Beware of using outdated weathering test methods simply because there is a historical pressure to pass a standardized weathering test. Insisting on passing such a test could require additional and mostly expensive light stabilisers, which would be necessary for meeting a nonsensical standard but not necessary for normal use. The outcome might be a new product whose price could rise unnecessarily.

Frequently, fluorescent UV lamp devices employing UV-B-313 lamps can still be found in testing laboratories for the purpose of monitoring current series production. Only if preliminary studies have sufficiently proved the existence of a connection between the weathering results of artificial weathering with a UV-B-313 lamp and the service life found in "normal use" could this lamp be further used in order to considerably shorten the artificial weathering - but only for the aforementioned purpose. In this case, if passing a fixed exposure period without any damage is taken as a criterion of sufficient coating quality, premature damage would signal a deviation from the normal production process. However, whether this deviation would also influence the coatings' ageing behaviour in normal use remains questionable.

6.2.2.5 Acceleration by increased irradiance

The radiation stress in artificial weathering can be increased up to a factor 10 relative to normal use if oriented toward the natural extreme of irradiance and if periods making little or no contribution to coating ageing are excluded. Since the time outlay is still too high, efforts are being made to increase the irradiation stress by increasing the irradiance as well. As described in detail in Chapter 3.1, the content of polymer free-radicals and oxygen molecules at the point where photooxidation takes place determines the rate of coating ageing. The concentration of polymer free-radicals depends mostly on the irradiance. However, the concentration of oxygen depends predominantly on the ability of the polymer to transport the oxygen molecules from the ambient air to the place of the reaction. In other words, the rate of oxygen diffusion is the second important factor that determines the ageing rate. Increasing the irradiance means producing a higher number of polymer free-radicals. As already described in Chapter 3.35, not all of the polymer free-radicals can be oxidized if oxygen and free-radicals do not come together in time. Otherwise, the polymer free-radicals will recombine with each other, mostly without causing any ageing. This is the risk of conducting artificial weathering with increased irradiance.

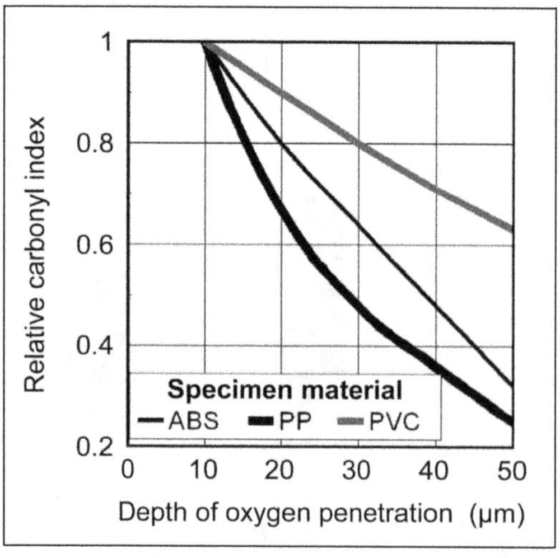

Figure 6.19: Typical oxidation profiles, of different weathered polymers, in the form of the carbonyl index as a function of penetration depth, starting from the irradiated surface and extending to a depth of interest for coatings in general (adapted from [12])

Figure 6.19 shows the relative carbonyl index as a function of the oxygen's penetration depth for different polymers. Due to the limited rate of oxygen diffusion, the concentration of oxidation products decreases with increasing film thickness, revealing the typical oxidation profiles. The profiles were measured by means of FTIR spectroscopy and microtome technology. The steep slope of the curves clearly shows the influence of limited oxygen diffusion.

Due to the competition between formation and recombination of the polymer free-radicals a regressive relationship between irradiance and rate of coating ageing can be expected. In other words, **doubling the irradiance need not lead to a doubling of ageing**. As already shown in Figure 3.5 for the example of PE the irradiance E_e acting on the specimen surface is given by the exponent 0.5. Figure 6.20 shows the expected regressive relationship between UV irradiance and polymer ageing for two different types of polyurethane artificially weathered in a xenon lamp device. The radiant exposure needed to effect a change in colour of $\Delta E^* = 2$ served as ageing criterion. In this example, enhancing the irradiance by a factor of 3 did not lead to a threefold decrease in exposure period needed for achieving the same ageing effect.

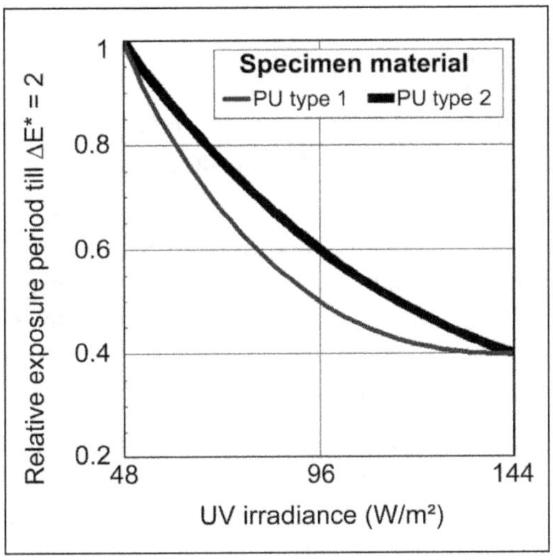

Figure 6.20: Exposure period needed to effect the same colour change in $\Delta E^* = 2$ as a function of UV irradiance for artificial weathering of two different polyurethane samples by means of a xenon lamp weathering device, adapted from [63]

To promote the use of unnaturally enhanced irradiance, the so-called law of reciprocity is often referred to in the literature [64]. It implies that the ageing effect will always be the same if the product of irradiance and exposure period also remains the same, irrespective of whether weathering is performed at high irradiance and short exposure or low irradiance and long exposure. Naturally, the coating surface micro climate also must remain constant. In artificial weathering practice, such a relationship can only be expected in a severely limited range of irradiance and in special cases.

Since the number of recombining polymer free-radicals always increases with increasing irradiance, an exponent of less than 1 must always be expected in the context of irradiance and rate of ageing, in the weathering of coatings, too. Exponents deviating not so extensively from 1 can be expected only if the outermost micrometers of the coating surface are responsible for essential effects of coating ageing, such as loss of gloss and change in colour. This is the case where the essential weathering stresses, such as oxygen and water are available direct from the ambient atmosphere without the need for travelling substantial distances by diffusion. If suitable analytical methods, such as FTIR spectroscopy and microtome technology, are available to determine the increase in oxidation products in the outermost micrometers of the coating surface, a better approximation to the law of reciprocity seems to be possible.

Exponents deviating more extensively from 1 can be expected in the weathering of basecoat/clearcoat systems where yellowing of the clearcoat or delamination of the coating layers occur by way of ageing effects. In this case, acceleration factors can be expected which are heavily dependent on the diffusion behaviour of the coating being tested, the chosen ageing criterion, and the change in coating surface micro climate due to unnaturally enhanced irradiance.

As discussed in Chapter 3.2.5, artificial weathering with unnaturally enhanced irradiance is not advisable for testing multilayer coatings. The thermal stress deviates too much from that which occurs in normal use, especially at the interfacial areas between the different coating layers. The use of enhanced irradiance in xenon lamp devices must be rejected unless additional means for reducing the correspondingly enhanced thermal load caused by VIS and NIR is installed. Such a case was reported in [103]. Two series of 33 different basecoat/clearcoat systems for automotive OEM were artificially weathered in a xenon lamp device by method X_{1a} at different irradiance. The one series was weathered at UV irradiance (300 to 400 nm) of 65 W/m², and the other of 130 W/m². The ageing response of the coatings to doubling of the irradiance differed. No ranking correlation between the two coating series was found for the loss of gloss (PEARSON = -0.06) or the change in colour (PEARSON = -0.06). For the same UV radiant exposure (about 312 MJ/m^2), weathering at different irradiance produced different ageing. In this case, the series exposed to the higher irradiance experienced less loss of gloss for almost all coatings. This example clearly proves that the **law of reciprocity would appear not to hold for modern automotive coatings.**

Similar to the effect of increased photon energy due to the use of lamps emitting unnaturally short wavelength, the **use of unnaturally enhanced irradiance will also lead to rankings in weatherability which deviate from those arising in normal use** if the ageing behaviour of different coatings is to be compared. With regard to the use of unnaturally enhanced irradiance during artificial weathering for the purpose of monitoring current series production, the statement which applies to unnaturally short wavelength holds again.

Finally, let us consider the technical possibility of generating enhanced irradiance for the purpose of artificial weathering. The first aim of developing weathering devices was to imitate global radiation, including global UV, so as to achieve similar spectral distribution and the same maximum irradiance. The first xenon lamp devices that went on sale provided irradiance of up to 1100 W/m^2. If the weathering standards in force at that time recommended lower irradiance, the devices were fitted with special accessories for reducing it during weathering. Modern weathering standards allow the use of irradiance three times as high as the maximum natural UV, i.e. up to 180 W/m^2 (300 to 400 nm). The corresponding xenon lamp devices are currently available.

However, care is advised with unnaturally enhanced irradiance. As the aforementioned examples revealed, the natural radiation stress should be simulated on the basis of the standardised curve as per CIE 85 Table 4 for weathering coatings. The use of higher irradiance can only be recommended for simulating the thermal stress on internal plastic parts of automobiles exposed to solar radiation that has passed through the car windows. Fluorescent UV lamp devices are much more limited in terms of irradiance for technical reasons. Their radiance can only reach 1.5-times the value of maximum natural irradiance

6.2.2.6 Continuous or alternating irradiation

As already presented in simplified form in Figure 3.2 and discussed in detail in Chapter 3.1.2, photooxidation of polymers initiated by absorbed UV photons commonly proceeds via a free-radical chain mechanism starting with the formation of polymer free-radicals. Figure 3.2 shows that once this chain mechanism has been initiated, it can be sustained without further absorption of UV photons and the formation of new polymer free-radicals. This is expressed by the reaction steps 2 and 3 in Figure 3.2. Interruption of the chain-propagation mechanism only occurs if the concerned polymer free-radicals are deactivated by reaction with each other or with the free-radical scavenger (see Chapter 3.1.4) to form neutral polymer molecules. A free-radical chain which propagates without the need for further irradiation is commonly called a dark reaction. As discussed in Chapter 3.1.2, once initiated, the dark reaction could theoretically lead to complete destruction of the polymer. Older literature, especially, attached great significance to the dark reaction and this was often allowed for by including dark phases in the test cycles of standardized artificial weathering methods. As also discussed in Chapter 6.2.1, dark phases have no significance in the weathering of modern, high-quality coatings due to the common use of highly effective light stabilisers. In particular, a content of sufficient free-radical scavengers makes the dark reaction subside in a very short time. Recent experience shows that **dark phases are increasingly seen as wasted time in artificial weathering**. Nowadays, avoiding dark phases is seen to provide major potential for time-compressing accelerated testing.

Figure 6.21 shows how shortening the dark phases in the test methods F_{2a} and F_{2b} from 12 to 8 hours a day shortens the exposure period needed for inducing crack formation in two differently stabilised samples of a waterborne acrylate/melamine clearcoat.

Modern weathering standards therefore increasingly recommend weathering cycles with test panels irradiated continuously by xenon lamp and fluorescent UV lamp weathering devices. The use of dark phases as a means of saving on the electrical running costs of fluorescent UV lamp devices is of no significance now. However, the use of dark phases to save on the electrical running costs of xenon lamp devices is worth consideration. Operating the xenon lamps with short light pulses alternating with short dark phases could be a suitable way of considerably shortening the consumption of electrical energy.

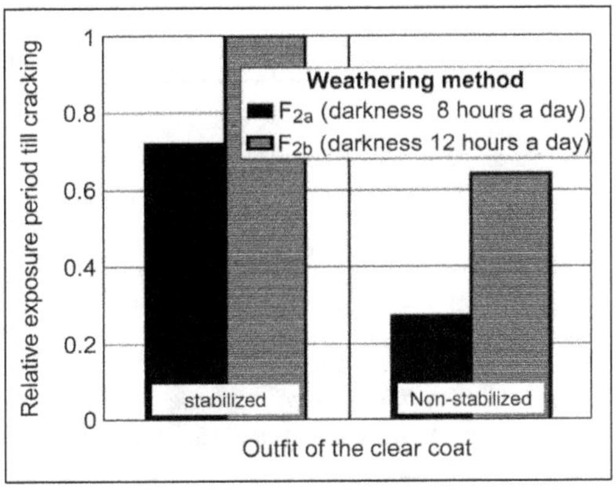

Figure 6.21: Relative exposure period to time of cracking as a function of the duration of dark phases during artificial weathering in a device employing fluorescent UV-B-313 UV lamps for a waterborne acrylate/melamine coating, adapted from [16]

6.2.2.7 Radiation measurement and control in weathering devices

An essential advantage of artificial weathering is the possibility of controlling the radiation stress within fixed values of irradiance. Appropriate measuring technology is needed if this advantage is to be exploited. This is mostly supplied with the purchased weathering device. Although automatic controlling is predominantly used in modern weathering devices, manual controlling is still in use. With regard to modern state of the art, lamps which are controlled only on the basis of hours of illumination without the actual irradiance being tracked no longer fulfil the requirements of modern test laboratories and good laboratory practice as demanded by ISO 17025.

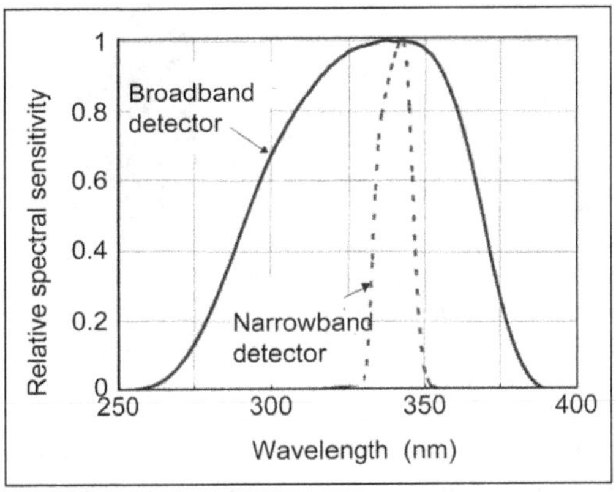

Figure 6.22: Spectral sensitivity of commonly used detectors for measuring and controlling UV irradiance in weathering devices, from [77]

It has proved expedient to track at least one parameter of UV irradiation (irradiance or radiant exposure) during artificial weathering either continuously or regularly. Two different methods have evolved in the case of xenon lamp devices. UV irradiance is controlled either by tracking the broadband or the narrowband values. Broadband means the whole range of UV (e.g. 50 W/m² from 300 to 400 nm) and narrowband means the spectral irradiance at a certain wavelength inside the UV (e.g. 055 W/m²/nm at 340 nm). Some standards recommend values for both. The reason for these two different concepts are mainly historical. Some weathering devices use broadband, others use narrowband and yet others use both concepts of measuring and controlling. Both technologies can be seen as equivalent from technical or economics aspects. Figure 6.22 shows the spectral measuring sensitivity of radiation detectors commonly used in modern UV radiometers

If the spectral distribution of the lamp or the lamp/filter system is known, broadband and narrowband values can be converted into each other. Narrowband technology is mostly used for xenon lamp devices employing irradiance at 340 nm. Fluorescent UV lamp devices mostly use broadband technology because their spectral distribution deviates more from that of the global radiation. Although both technologies are almost equivalent, broadband technology should be preferred for controlling irradiation stress. In order to decrease the uncertainty of the weathering test, the use of both technologies is recommended for the purpose of tracking and recording

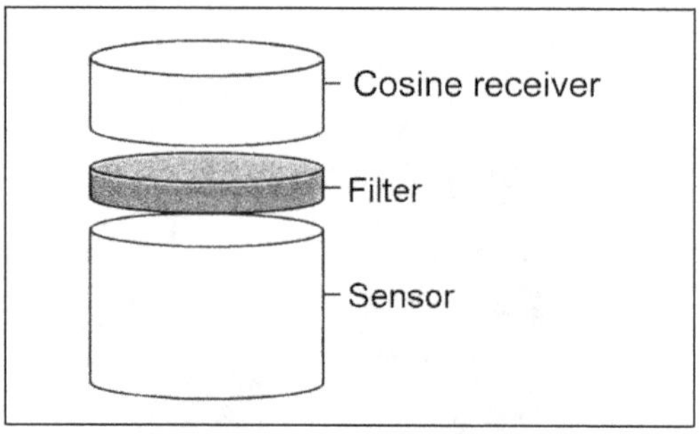

Figure 6.23: Simplified design principle of a UV radiometer commonly used in artificial weathering

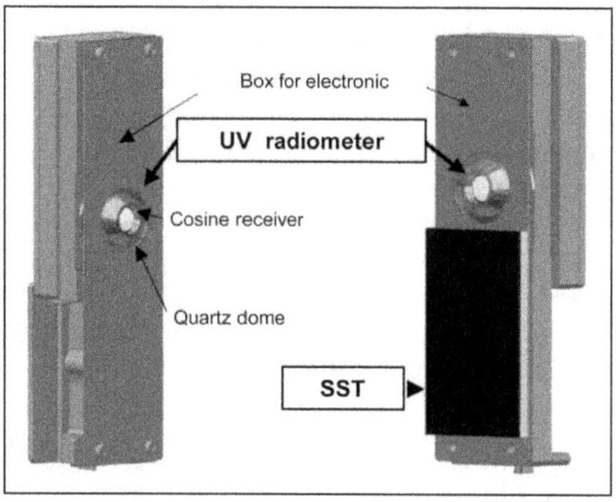

Figure 6.24: Modern UV radiometers for artificial weathering, as stand-alone version (left) or combination with a black standard thermometer (right), adapted from [77]

the UV irradiance during the test. Given that ISO 17025 demands high conformity in testing and measuring, agreement on the technology would be desirable. Nowadays, high-quality instruments are available that record the chosen controlled parameter. Their spectral sensitivity is commonly effected by special optical filter systems. The basic design of a typical UV radiometer for use in weathering devices is presented in Figure 6.22.

On the top of the radiometer is the cosine receiver which is an important component. Its purpose is to receive all the radiation emanating from the half space above the radiometer. Low-quality cosine receivers are a common source of mistakes where measuring of irradiance is concerned. This is particularly critical if the radiometer is used as a standard for calibration of other or secondary instruments. Sometimes, low-quality radiometers have no cosine receivers at all. Other radiometers, even high-quality types, display remarkable differences in the correctness of the cosine receiver design. Equipping weathering devices with cosine receivers of different performance will inevitably lead to different irradiation stress. Even identical weathering devices can yield different results. This is one of the reasons for the poor reproducibility of artificial weathering tests carried out in identical weathering devices but operated in different testing laboratories. The cosine receivers are commonly made from unselectively scattering panes of quartz or Teflon. Figure 6.24 shows photographs of two typical versions of UV radiometers for continuous use in xenon lamp devices. The picture on the left side shows a stand-alone radiometer whereas that on the right side shows a radiometer combined with a black standard thermometer for registering the surface temperature of the test panels.

They are designed for easy mounting at the same level as the test panels. Note that the influence of the temperature on the measured values of irradiance must not be neglected. Calibration must therefore be performed for the whole temperature range used for the weathering tests. High-quality radiometers are frequently equipped with means of automatic temperature compensation. Note that radiometers have to be calibrated separately if used in weathering devices equipped with different types of lamps. As reported in [78], broadband radiometers calibrated for use in fluorescent UV lamp devices indicated false UV irradiance values when used in xenon lamp devices.

In summary, it can be said that all types of radiometers presently used for calibration are suitable only for measuring the quantity of irradiation stress. They are unsuitable for measuring the quality of irradiation, i.e. their spectral distribution. This property of the radiation is known to heavily influence the ageing behaviour of the coatings to be tested. Measuring the spectral distribution of the radiation received by the test panel surface requires expensive spectroradiometers. Their use in common testing laboratories cannot be economically justified. This underscores the need for cheaper ways of monitoring spectral distribution, e.g. by means of corresponding reference coatings.

6.2.2.8 Monitoring of irradiation by means of reference coatings

Figure 6.25 shows a typical testing laboratory that uses a lot of weathering devices, as is often found in the paint industry. As can be seen in this photograph, a large number of mostly identical devices are employed in the laboratory's testing area.

Slightly darkened panes make it possible to look into the weathering devices. The luminosity perceived by the observer suggests that each of the lamps inside the devices produces light of the same intensity. Displays that can always be found on weathering devices signal continuous readings of test parameters, such as temperature, relative humidity, irradiance and others in bright digits. Going from one weathering device to the next, the mostly good agreement between the displayed values is remarkable. These indicate that the actual testing parameters also agree to within several decimal points. Given the discussions on repeatability and reproducibility of weathering in general, however, doubts arise as to the conformity of the weathering stress from device to device and from test to test in spite of the fact that the readings for the test parameters are identical. Experience with coatings testing supports such doubts. Considerable uncertainty in the observance of fixed test parameters exists for each artificial weathering device and therefore must not be underestimated. An assessment of this uncertainty for compliance with standardized or otherwise fixed test parameters is also demanded by ISO 17025 [53] as an important quality-criterion of weathering tests. The test engineer, whether working for paint makers or coating users, has to prove that the test has been done correctly. However, equipping weathering devices with additional or more accurate measuring technology quickly reaches the limits of what is economical.

On the other hand, not even extended measurement technology can remove all doubts that the parameters are being correctly observed. For this reason, it is now standard practice to use reference coatings to reduce uncertainty in artificial weathering, too. Especially **in coatings testing the use of reference materials in artificial weathering is one of the basic requirements to be met by a testing laboratory seeking accreditation according to ISO 17025.**

As already described in Chapter 4.4 in the context of outdoor weathering, reference materials in artificial weathering serve the same purpose, i.e. monitoring the weathering stress during the test by simple technical means, additionally to and independently of, instrumental means of determining individual test parameters. As also described in Chapter 4.4, reference materials should have the identical ageing behaviour as the coatings to be tested. In other words, coatings should predominantly serve as reference materials and be preferred to other types of polymeric materials. Reference coatings must also lend themselves to being handled and assessed in the same way as the coatings to be tested. As already stated in Chapter 4.4, standardised and internationally accepted or certified reference coatings for monitoring compliance of fixed weathering stresses in coatings testing are not currently available. Nor are there any technical solutions for tracking individual weathering stresses, such as irradiance, heat or moisture, in isolation from each other by means of reference coatings.

Figure 6.25: View of a typical testing laboratory for artificial weathering

For coating weathering performed outdoors or artificially, the only technical solution which is currently suitable as a means of monitoring entails the use of a more or less high number of coatings of known ageing behaviour along with the coatings to be tested. **Reference coatings are best selected by the user of the weathering service from his own samples.** Such coatings, the user's own reference materials, should differ in ageing behaviour as much as possible and these needs to have been demonstrated in previous artificial weathering tests. Of course, the general requirements for a reference material concerning repeatability and reproducibility must also be fulfilled. Nevertheless, monitoring with such reference coatings only permits an assessment of compliance weathering stress as a whole, i.e. without the possibility of differentiating between the individual weathering stresses. Yet, deviations in the ageing behaviour of the reference coatings from the one test to the next can be caused by each of the various weathering stresses. In other words, in assessing the repeatability of a certain artificial weathering test, it is not possible to tell whether deviations in temperature, relative humidity, irradiation or all together, were responsible for differences that occurred in the ageing behaviour of the reference coatings. Thus, the general desire to be able **to assess the repeatability of the entire individual weathering stresses which occur in artificial weathering in isolation from each other by means of reference materials remains unfulfilled.**

Only the current possibility of monitoring irradiation in isolation from the other weathering stresses goes any way towards meeting this desire. Current reference materials at least allow the repeatability of irradiation stress to be assessed in isolation from that of the coating surface micro climate. This affords a way of approximating the conformity in testing demanded by ISO 17025. If this were applied to inter-laboratory comparison the reason for possible differences in weathering results might be found more easily. Therefore, given its particular significance to weathering in general and considering the technical possibilities, separate monitoring of irradiation appears sensible and feasible. However, monitoring irradiance in artificial weathering is not as easy as it is in outdoor weathering. Several monitoring functions prove necessary in artificial weathering. They are listed in Table 6.7.

As described in Table 6.7, clear differences from outdoor weathering arise. This was the reason that reference materials for outdoor und artificial weathering are treated in separate chapters in this book. The differences in the necessary extent of monitoring are due to different sources of radiation. While the sun can be seen as an invariable radiation source, artificial radiation sources either differ considerably or change in the quantity and quality of radiation which they emit. For this reason, additional monitoring functions are necessary in artificial weathering. Due to the properties of the sun, monitoring functions 2 and 3 in Table 6.7 are not necessary for outdoor weathering. In contrast to the sun, artificial radiation sources in the form of xenon lamps and fluorescent UV lamps suffer from ageing due to their own radiation and this leads to changes in quantitative and qualitative irradiation parameters.

Table 6.7: Monitoring functions that must be realised for assessing the repeatability of irradiation stress during artificial weathering

Monitoring function	Technical solution	Integration-period	Suitable reference material	Necessary properties
1. UV radiant exposure 300 ... 400 nm (uniformly distributed over the whole target board)	summation of the effect of the whole radiation >300 nm	if possible the total weathering test	ORWET reference coating	low UV selectivity. low temperature and moisture sensitivity
2. relation of UV radiant exposure <350 nm to that > 350 nm	summation of the effect of the UV radiation at (320 ±20) nm	if possiblethe total weathering test	polystyrene-chips special inks solved in water	high UV selectivity, low temperature and moisture sensitivity
3. sure avoidance of UV radiation <300 nm	summation of the effect of the UV radiation <300 nm	Mostly limited (e.g. 180 MJ/m² using BASF-ref.)	BASF-clear coat crack standard	high UV selectivity

Lamp ageing is the reason that the lamps regularly need to be replaced after 1000 to 2000 hours in use. The optical filters needed for providing the fixed spectral distribution undergo a similar kind of ageing, too. Filter systems in xenon lamp devices must be replaced after 4000 hours usually. Lamp ageing manifests itself as decreasing irradiance. Ageing of lamps and filter systems used in xenon lamp devices additionally leads to qualitative changes in irradiation. Qualitative change here means shifting of the curve of the UV range to a higher wavelength. This produces a substantial decrease in UV irradiance. UV, which is the most effective wavelength range for coating ageing, gradually decreases with respect to VIS and NIR. This ageing-induced change in radiation spectrum cannot be seen either by measuring the narrowband irradiance at 340 nm or by measuring broadband irradiance of the whole UV from 300 to 400 nm. Therefore, monitoring the irradiance on the basis of this kind of measurement, as is commonly practice, cannot undo the altered spectral distribution inside the UV wavelength range. This kind of monitoring can only offset the decrease in the whole UV irradiance, but will not repair the radiation spectrum, i.e. the ratio of irradiance below 340 nm to that above 340 nm.

Fluorescent UV lamps are less known for undergoing a change in radiation spectrum up to the end of their service life, but this only applies to devices fitted with one type of lamp, e.g. UV-A-340. However, lamp ageing must be considered if combinations of fluorescent UV lamps of different radiation spectra are used, as is recommended in EN 4892-3. As observed, different types of lamps age at different rates. Consequently, radiation emitted by a lamp combination can undergo a marked change in radiation spectrum.

Besides the ageing of lamps and filters, deviations from the standardized or otherwise fixed radiation spectrum can also be caused by the use of wrong or damaged lamps or filters or certain subjective influences. If the radiation spectrum is not tracked or monitored during the weathering test or at least from test to test, poor repeatability of the weathering results must be expected.

As also observed, variations in irradiation stress can also be caused by differences in certain features of the test panels and the test chamber, such as the extent of the target area occupied by test panels, local distribution of the test panels inside the test chamber, the reflectance of the test panel surfaces and that of the chamber walls. As measured by [63] during weathering in a xenon lamp device completely filled with differently coloured test panels, the ratio of the spectral irradiance at 320 nm to that at 370 nm varied by up to 12 %, the precise figure depending on the position where the UV radiometer was mounted. Furthermore, limited local homogeneity in irradiance over the available area of the target board must always be expected due to technical and economical aspects. For this reason, weathering devices will still be seen as norm-compliant if the irradiance deviates locally from the maximum value by up to 20 %. However, only a fluctuation of up to 10 % from the maximum is commonly seen as sufficiently homogeneous. In order to minimise radiation-induced fluctuation of the weathering results, it is common test practice to regularly swap the positions of the test panels on the target board. Evidence to support whether such regular changes meet the requirements for local homogeneity of irradiance can only be adduced at high technical outlay.

A further important aspect of monitoring irradiation stress is the diversity of lamps, filter systems and weathering devices presently used in coatings testing. The staff involved in coatings testing have to make sure that the cut-on of the UV radiation corresponds to the fixed test specifications. As has been mentioned several times in this book, the position of the cut-on strongly influences coating ageing. If the cut-on lies below the natural value of 300 nm, ageing effects can arise that never occur in "normal use". In contrast, if the cut-on is at too high a wavelength, important ageing effects will not take place. This may lead to an ageing rate that is lower than in normal use.

As described in Chapter 6.2.2.2, xenon lamps are always used in combination with filter systems. At the moment, filter systems are still being widely used in coatings testing whose cut-on lies below 300 nm (e.g. quartz/boro filter system). Also, the UV-B-313 fluorescent UV lamp, which emits

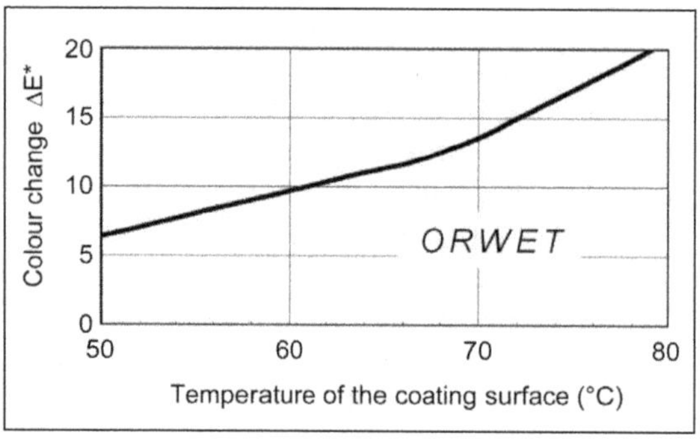

Figure 6.26: Influence of temperature on the colour change of an ORWET reference coating during artificial weathering in a xenon lamp weathering device after 40 MJ/m² of radiant exposure, adapted from [80]

radiation below the natural cut-on, is still used in coatings testing. Especially where testing laboratories using several different types of fluorescent UV lamps or different types of filter systems, mix-ups between lamps and filters during regular replacement cannot be completely ruled out. Note that, in certain testing laboratories, weathering is only one of their tasks and the staff's specialist knowledge may be rather limited. Lamp and filter replacement is often done by the staff of the testing laboratory itself without their being aware of the consequences of possible mistakes. For instance, the presence of a relatively small amount of radiation below 300 nm cannot be detected with the measuring instruments commonly found in a laboratory for coatings testing. However, even this small, hard-to-detect wavelength range has an enormous influence on coating ageing.

Based on the latest evidence, all necessary monitoring functions listed in Table 6.7 can be realized at reasonable costs and technical outlay. Commercially available reference materials can be used that have already proved appropriate for this purpose. They therefore will also make suitable reference materials if the laboratory performs regular internal quality controls or takes part in inter-laboratory comparisons or proficiency-testing programs, as demanded by ISO 17025 [79]. As set out in Table 6.7, the monitoring functions will now be discussed individually.

Monitoring function 1 serves to prove that the coatings to be tested were exposed to the fixed UV radiant exposure in the whole wavelength range from 300 to 400 nm during the weathering test. Mounting at different positions on the target board function 1 also provides evidence that all test panels received identical radiant exposure. This function can be realized by means of the ORWET reference coating, which also serves as reference material for outdoor weathering and is described in Chapter 4.4. ORWET is quite cheap, easy to handle, does not interfere with the test and does not overly restrict the area available for testing. As described by [80] and presented in Figure 6.21, the influence of the temperature on the change in colour is considerably less than would be expected from Arrhenius' law. Fluctuations in temperature can therefore be virtually neglected, at least within the usual tolerance for fluctuation from one artificial weathering test to the next. For instance, a fluctuation of ±3 °C should lead to a fluctuation in relative colour change of just ±10 %.

The same considerations apply more or less to the influence of moisture. However, it must be ensured that the radiation-induced change in colour is limited to the range of linear correlation with radiant exposure.

Monitoring function 2 (Table 6.7) serves to prove that the coatings to be tested were exposed to the fixed UV radiant exposure over the full wavelength range from 350 to 350 nm during the weathering test. This means that the position of the slope of the UV radiation did not change from test to test. This function can be achieved only by introducing a second reference material which is not overly sensitive to radiation of wavelength <320 nm. Its spectral sensitivity at 320 nm must at least be 10 times that at 370 nm. This is the most essential difference from ORWET, which has very poor spectral selectivity. A coating that adequately meets the aforementioned requirements is not yet available. However, an appropriate substitute is available in the Polystyrene Lightfastness Standard of SAE J1960 [81]. Although the chips employed therein are not a coating, they possess the key properties of a reference material.

As described in [80], the commonly employed three-millimetre-thick chips made from clear polystyrene are designed to be reproducible from one batch to the next and have served as reference weathering material for exterior automotive applications for years. Yellowing of the chip occurs linearly with increasing radiant exposure. As with ORWET, there is a limitation on the linear correlation as yellowing approaches the saturation level. Easy to handle and with an ageing behaviour which is easy to monitor, the polystyrene possesses the specified selectivity in the short-wave UV range. Figure 6.27 shows its spectral sensitivity in the wavelength range from 300 to 340 nm, in the form of the radiant exposure necessary for reaching identical yellowing. As may be seen, its spectral sensitivity drops rapidly with decreasing wavelength. Upon irradiation at 340 nm, the polystyrene chip needs 20 times less exposure than for irradiation at 340 nm. In comparison, and under the same conditions, ORWET needs only 2.5 times less time due to its smaller spectral selectivity.

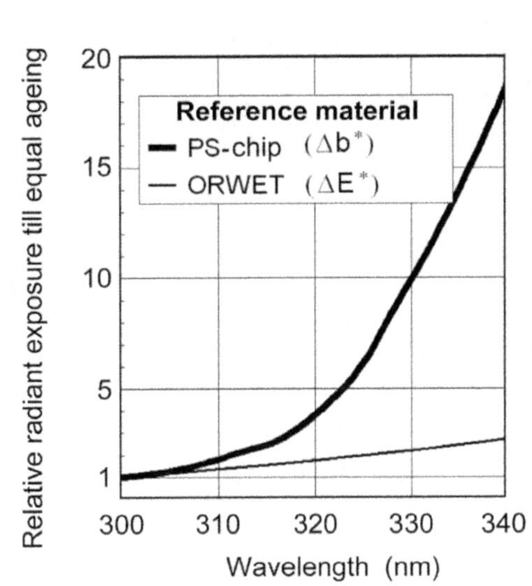

Figure 6.27: Spectral sensitivity of standard polystyrene lightfastness chips relative to that of the ORWET reference coating, demonstrated for the relative radiant exposure required for effecting an identical colour change in the wavelength range 300 to 340 nm, adapted from [42]

As reported in [80], virtually identical statements can be made about the temperature influences and moisture as for the reference coating ORWET. Due to its high transparency, it is important to cover the back of the polystyrene chips with a metal sheet, preferably of stainless steel, in order to prevent reflected radiation from passing through the polystyrene a second time.

As described in [79], certain inks dissolved in water and filled in small transparent plastic tubes respond to radiation below 320 nm with a strong colour change and also fulfil the requirement of reference material for this monitoring function.

Monitoring function 3 (Table 6.7) serves to demonstrate the absence of radiation of wavelength below 300 nm. This function needs a third reference material of the right selective spectral sensitivity. The “BASF clearcoat crack standard” conforming to VDA test specification 621-430” meets this requirement. It is a special acrylate/melamine-based clearcoat applied to a silver-metallic basecoat. Steel panels measuring 10 x 10 mm commonly serve as substrate. This reference coating has also been used for testing automotive coatings for years, mostly in weathering devices in which the test panels are irradiated with a considerable amount of wavelengths below 300 nm. As described in [82], the “BASF clearcoat crack standard” responds with the sudden formation of cracks when a certain radiant exposure is reached. In comparison to other methods of coating weathering, the “BASF clearcoat crack standard” reaches its ageing criterion (crack formation) after a relatively short exposure period in xenon lamp devices equipped with quartz/boro filters and in fluorescent UV lamp devices employing UV-B-313 lamps. In contrast, extensive studies of the ageing behaviour of the “BASF clearcoat crack standard” showed that weathering in devices without radiation below 300 nm as well as outdoors in Florida did not lead to formation of cracks, even though artificial weathering tests with UV radiant exposure up to 270 MJ/m², equivalent to 1500 hours’ exposure, and outdoor weathering up to 2 years were involved.

So a reference coating was found that undergoes crack formation in response to unnaturally high photon energy only. Determination of the spectral sensitivity of the “BASF clearcoat crack stand-

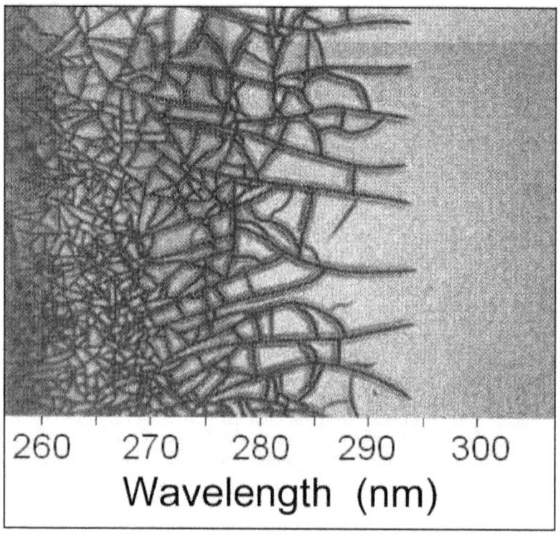

Figure 6.28: Spectral sensitivity of the "BASF clearcoat crack standard" spectrally irradiated with monochrome radiation, whereby crack formation served as the ageing criterion, adapted from [83]

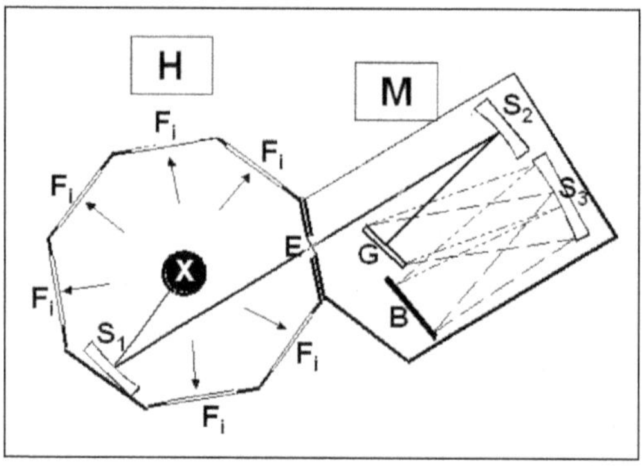

Figure 6.29: Apparatus for determining the spectral sensitivity of polymers, adapted from [42]

ard" confirmed this statement. The spectral sensitivity was determined spectrographically by means of a monochromator, as described in more detail in Chapter 6.2.2.9. Crack formation as a function of the wavelength of the monochrome radiation is presented in Figure 6.28. It shows the photo of the "BASF clearcoat crack standard" monochromatically irradiated in the most interesting wavelength range below 300 nm. The start of crack formation was unequivocally assigned to a wavelength of 295 nm. Thus, the lack of cracks on the "BASF clearcoat crack standard" after artificial weathering can be viewed as evidence that no radiation below 300 nm was present during the test.

6.2.2.9 Determination of the coating's spectral sensitivity

Economic aspects, as discussed in Chapter 6.2.2.3, create a tendency for the test engineer to prefer fluorescent UV lamps in answer the question as to "which type of lamp is best suited to my task?" Knowledge of the spectral sensitivity of the coatings to be tested may help to decide between a xenon lamp and a fluorescent UV lamp. As described in Chapter 3.2.1.2, spectral sensitivity is a pure coating property. Its curve gives the test engineer the necessary information as to whether only global UV, which permits the use of the economically advantageous fluorescent UV lamp, or also further parts of the global radiation (e.g. VIS), which requires the economically disadvantageous lamp, must be simulated. If the spectral sensitivity proves that VIS does not cause any marked ageing effect, there is no need to simulate it. However, the other criteria for deciding between a xenon lamp and a fluorescent UV lamp, as discussed in Chapter 6.2.2.3, must be taken into account. Knowledge of the spectral sensitivity is also of interest to the paint maker, providing help with the creation or selection of the right light stabilisers.

Two main methods have commonly been employed for determining the spectral sensitivity of polymers as described in [42]. These are the spectrographic technique and the sharp cut-on filter technique. The spectrographic technique is the more elegant method. In both methods, the specimens were irradiated with virtually monochromatic radiation. This can be either the radiation spectrally dispersed by a monochromator or the radiation filtered by a set of appropriate cut-on filters. An apparatus capable of performing both methods is operated in the Federal Institute for Material Research and Testing (BAM) in Berlin, Germany. It is shown schematically in Figure 6.29 and explained in the following text.

A 900 W xenon short arc lamp, labelled X, serves as radiation source and is centrically mounted inside an approximately spherical lamp housing, H, of about 35 cm in diameter.

First, let us look at the **spectrographic technique**. In this case, the radiation emitted by the lamp is focused on the entrance slit, E, of a highly luminous 3 cm-monochromator by means of the spherical mirror, S. A second spherical mirror, S_2, mounted inside the monochromator box, M, reflects the radiation onto the optical diffraction grating, G. This grating disperses the radiation linearly. By means of the concave mirror, S_3, the spectrally dispersed radiation is focused onto the target board, B, where the test panel of the coating to be tested is mounted. The main advantage of the spectrographic method is its high spectral resolution, which reaches about 4 nm per 1 mm along the irradiated test panel surface. After irradiation, the lateral distribution of the chosen ageing criterion along the spectrally irradiated test panel is determined. In other words, only laterally measurable properties can be studied a fact which can be seen as a limitation of the method.

Now, let us look at the **sharp cut-on filter technique**. It can be performed with the same apparatus. For this purpose, a corresponding set of spectrally different sharp cut-on filters, F_i, is mounted like windows in round openings in the envelope of the lamp housing, H. All filters are equidistant from the lamp, X. Figure 6.30 shows the different optical behaviour of a set of eight commercial filters (F_1 to F_8) often used for this method. They differ in the position of the slope and cut-on of their spectral transmittance. The test panels to be irradiated are mounted directly and simultaneously irradiated behind the filters. The differences in the measurable exposure results are attributed to the variations in the spectral radiant exposure. The disadvantages of this method are, first, the relatively broad spectral ranges yielding the information and, second, the limited number of irradiation places that can be used simultaneously. This leads to considerably lower spectral selectivity relative to the spectrographic technique. Bear in mind that the spectral transmittance of optical filters changes on ageing and may vary with temperature.

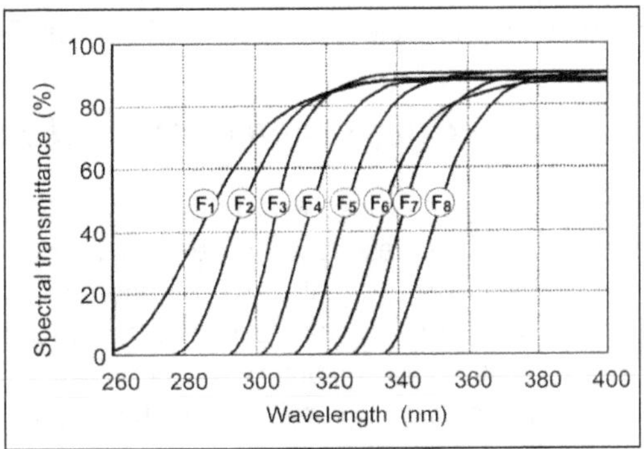

Figure 6.30: Transmittance curves for eight different optical cut-on filters (F_1 to F_8) suitable for determining spectral sensitivity, adapted from [42]

Since cut-on filters limit the radiation spectrum of the shorter wavelength only, the results of both techniques can be used jointly to assess whether synergies influencing coating ageing are at work between UV and VIS or between UV < 350 nm and UV > 350 nm, as described in Chapter 6.2.2.3. This information can also help in the decision between xenon lamp and fluorescent UV lamp.

Figure 6.31 shows the result of the sharp cut-on filter technique for the example of the reference coating ORWET. In this case, five cut-on filters were used. ORWET is a typical example which proves that even radiation above 470 nm can contribute to the change in colour, as already evidenced in Figure 6.5 by the use of the spectrographic technique.

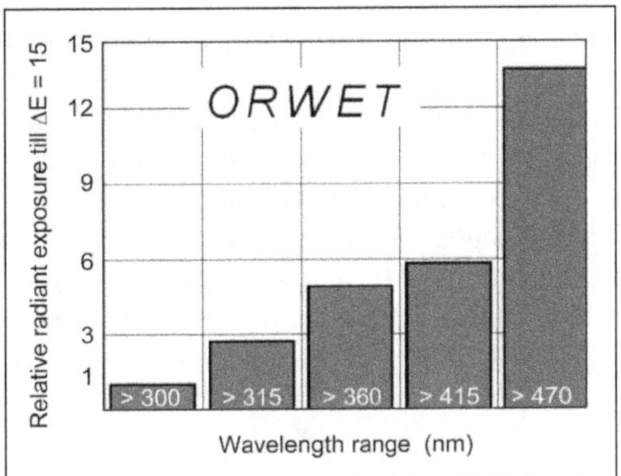

Figure 6.31: Radiant exposure necessary for effecting an identical colour change in the ORWET reference coating by means of xenon arc radiation whose cut-on point has been adjusted with a series of cut-on filters, from [71]

The form of presentation chosen for Figure 6.31 also clearly reveals (see the bar for the filter at >415) that, in this case, irradiation only with visible light (VIS) needs about 6-times as much radiant exposure as irradiation with the total xenon arc radiation down to 300 nm if identical changes in colour are to be attained. When viewed in conjunction with Figure 6.5, there can be no synergies, as mentioned before. In other words, the weatherability of ORWET coatings can be tested in fluorescent UV lamp devices.

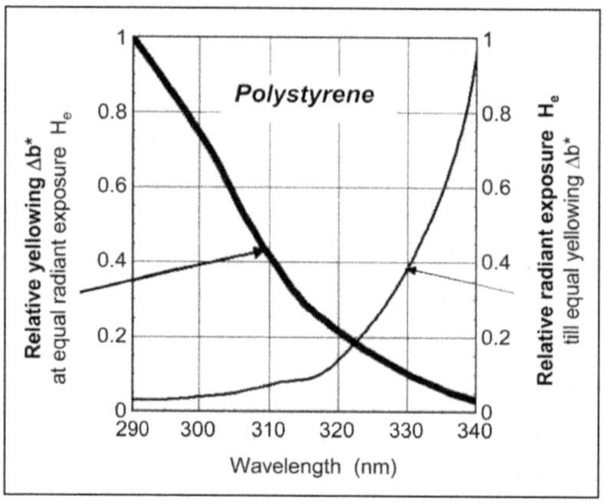

Figure 6.32: Ways to illustrate spectral sensitivity for the example of weathering-induced yellowing Δb of polystyrene chips (widely used as a reference material) for the wavelength range 290 to 340 nm*

As shown in Figures 3.20, 4.28, 6.4, 6.5 and 6.27 and demonstrated in Figure 6.32 for the example of polystyrene reference material, spectral sensitivity can be presented in two different forms. Either the radiant exposure He (e.g. in MJ/m²/nm) necessary for achieving an identical ageing effect (e.g. change in colour $\Delta b^* = 2$) or the value of the ageing effect (e.g. Δb^*) reached at a fixed radiant exposure (e.g. 1 MJ/m²/nm) is plotted as a function of wavelength.

As can be easily seen, one presentation form is the reciprocal of the other. Note that spectral sensitivity is only valid for the tested polymer and only valid for the chosen ageing criterion. Considerable differences in spectral sensitivity can be expected for different coatings. Furthermore, consideration should be given to a possible influence exerted by the coating surface micro climate on the shape of the curves during irradiation. Spectral sensitivity determined in a normal room climate may deviate when transferred to real outdoor conditions.

6.2.3 Coating surface micro climate during artificial weathering

Good repeatability and reproducibility alike should be expected during artificial weathering according to the present state of the art. As mentioned in Chapter 6.2.2.8 however, it was reported that present testing practice with identical weathering test methods yielded unacceptable deviations in weathering results, even when carried out in identical weathering devices. Nowadays, the reason for such unsatisfactory reproducibility needs to be sought less in unexpected fluctuations in irradiation than in variations in the coating surface micro climate, i.e. in differences in thermal and water stresses. Modern possibilities for controlling irradiation stress, such as the use of regularly calibrated measuring instruments and concurrent weathering of reference materials as described in Chapter 6.2.2.8, corroborate this statement. However, deviations in the thermal and water stresses can occur in artificial weathering in undesirable and unnoticed ways.

First, consider the coating surface micro climate in xenon lamp devices. Although equipped with corresponding technical means of controlling the climatic test parameters in the chamber, such as ambient temperature, BPT/BST, ambient relative humidity, and TOW, the actual climatic situation on the coating surface itself cannot be measured directly. Coatings differing in surface properties or their position inside the weathering device can differ considerably in their coating surface micro climate. In other words, uncontrollable differences between the climate displayed in the test chamber and the climate actually acting on the surface of the test panels must be expected in xenon lamp devices. As is known, such differences exert a particular influence on coating ageing and therefore the weathering results. Overcoming this weakness of artificial weathering in xenon lamp devices is not possible at reasonable cost.

Irradiation with VIS and NIR decisively influences the surface micro climate of the coatings to be tested. In order to avoid an overly high surface temperature at a fixed UV irradiance, most of the absorbed radiation must be removed from the test panels in the form of heat by convection to the ambient air. This is commonly effected by a strong air stream directed onto the test panel surface, naturally not during phases of wetting with water. The changeable velocity of the air stream used for cooling the test panels commonly serves to control the temperature of the test panel surface. However, this is done based on BST/BPT, and not on the actual temperature of the coatings to be tested. In other words, white-coloured samples are cooled with the same intensity as black-coloured ones. Different surface temperatures cannot be ruled out in spite of identical BST/BPT. Achieving flow conditions which are locally homogeneous and temporally constant in a space of limited size such as a standard test chamber is well known in flow mechanics. Therefore, the design of modern weathering devices must be seen as compromises between ideal air flow conditions and testing capacity. That raises doubts about the homogeneity of the air velocity on all test panels mounted in the test chamber. Differences in surface temperature from one test panel to the next cannot be ruled out. **Controlling the surface temperature of the coatings to be tested on the basis of BST/BPT is not the same as controlling the actual surface temperature.**

Complementary information about the actual surface temperature can be gained by the additionally mounting a white standard thermometer in the xenon lamp device. The use of black and white thermometers enables the fluctuation of the coating's actual surface temperature to be assessed. For controlling the relative humidity in xenon lamp devices, it is standard practice to vary the water content of the air (often expressed by the dew point) used for cooling the test panels. Weathering devices are equipped with corresponding means of drying or wetting the test chamber atmosphere or wetting the test panel surface. The same statement which applies to surface temperature can be made for the relative humidity immediately acting on the coating surface. The relative humidity on the test panel surface changes inevitably with change in surface temperature. **Coatings that differ in their thermal stress differ in their water stress too when weathered in a xenon lamp device.**

What about the coating surface micro climate in fluorescent UV lamp devices? Since only UV radiation is used for weathering, the test panels weathered in a fluorescent UV lamp device experience only about 6 % irradiation stress relative to a xenon lamp device at identical UV irradiance. Therefore, the coating surface micro climate can roughly be seen as identical with the climate inside the test chamber. Consequently, controlling the coating surface micro climate is much easier in fluorescent UV lamp devices than in xenon lamp devices. Less uncertainty concerning compliance of the fixed weathering stresses can be expected in fluorescent UV lamp devices, although air flow-related local variations in the test climate must also be considered.

The following discussions of water stress apply to xenon lamps devices and fluorescent UV lamp devices alike. To make sure that, in accordance with the specified relative humidity, the test panels in the dry phases of the test cycles are only stressed by water in gaseous form, humidification of the chamber air should be performed, preferably by evaporation or steaming methods. Ultrasonic atomizing does not provide sufficiently aerosol-free air. Deposition of aerosol particles on the test panel surface does not fulfil the requirement for water stress in gaseous form and can contribute to poor reproducibility.

Two different methods are used for phased wetting of test panels in modern weathering devices. Spraying with water is the most widespread. The other method utilizes condensation of water on the test panel surface from a highly humid test chamber atmosphere. Remarkable differences in the influence exerted by these two wetting methods on coating ageing have been observed and reported by [48]. Spraying demands higher water purity to avoid influencing the test panels or parts of the weathering device, for instance, by limestone or salt crusts. The condensation method does not give rise to any problems in terms of the water's purity. However, it has been observed that the test panel surface is often incompletely or nonuniformly wetted. The water is induced to condense on the test panels' front faces by cooling their back faces. If the thermal conductivity of the test panels is too low, no condensation will occur. Therefore, wetting by means of condensation cannot be used for artificial weathering of coatings applied to plastics, concrete, plaster, ceramics or wood. Consequently, means such coatings cannot by weathered on their real substrate, as is generally recommended.

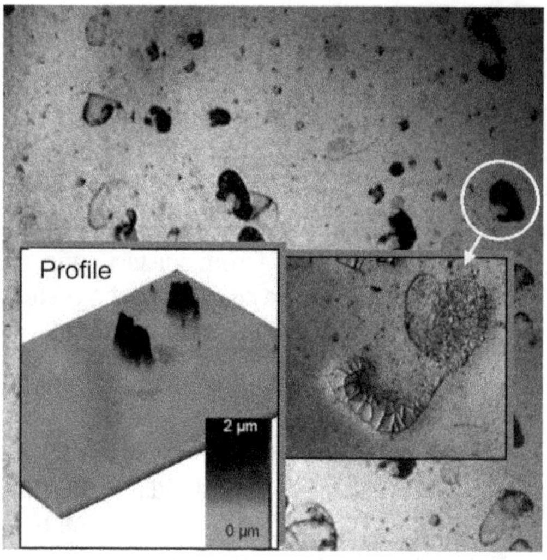

Figure 6.33: Photograph of a Chapter (12 mm x 12 mm) of the artificially weathered surface of a 2-pack PUR clearcoat containing numerous water spots (in the background), and a photograph (right fore-ground) and profile (left foreground) of a selected single spot at higher magnification

There is a further important disadvantage of the condensation method. It cannot be used to simulate a worst case weather scenario that is characterized by the mutual action of water and UV. As already discussed in more detail in Chapter 3.2.5, simultaneous application of liquid water stress and UV irradiation stress proved to offer major potential for accelerating artificial weathering. For this reason, the term "TOW+U" (time of wetness with UV) has been introduced in this book because of its particular influence on the rate of coating ageing. Insisting on weathering tests employing the condensation method, for instance, as recommended in ISO 4892-3, would mean wasting a third of the time available for irradiation.

Therefore, to achieve maximum time-compression or acceleration during artificial weathering of coatings, methods recommending continuous spraying simultaneously with UV irradiation should be preferred.

Note that the **test panel surface should always be covered with an intact water film during wetting!** This could also be achieved with a sheet of water on the coating surface. Note that all reaction partners involved in photooxidation, such as UV, water and oxygen, can act in concert in all phases of the weathering test. The formation of the free-radicals OH• and HO_2•, as shown in Figure 3.38, which have an accelerating effect on photooxidation, must not be impeded.

A further, oft-discussed problem is the purity of the water used for spraying. Economics aspects dictate that the water be circulated in a closed loop system. The exaggerated purity that is often specified in weathering standards, e.g. the use of de-mineralised, deionised or distilled water, does not seem to make much sense. Impurities or water-soluble degradation products released from the test panels during weathering can be expected to contaminate the water rather quickly. Sometimes, tap water is recommended in the literature. The water spots revealed on modern automotive coatings by artificial weathering with demineralised water raised the issue of purity recently. The background picture in Figure 6.33 shows a detail of the surface (12 mm x 12 mm) of a test panel weathered for 42 days in a fluorescent UV lamp device as per ISO 4892-3. The test cycle involved four hours of uninterrupted spraying a day.

The typical water spots formed during artificial weathering were clearly visible, and detracted from the optical impression of the coating surface. However, changes in gloss or haze were not measurable. A water spot about 1 mm in length was chosen and is presented at higher magnification in the right foreground. The profile (left foreground) revealed it to be material protruding about 2 µm above the level of the surrounding, undamaged coating surface. Identical or similar appearances of coating ageing were not found on test panels of the same coating during outdoor weathering. This proved that the dirty rain water was unable to induce formation of water spots. Furthermore, artificial weathering by means of the ADF test described in Chapter 6.2.4 in which the test panels were regularly exposed to water with a low content of acids failed to reveal any water spots. Chemical analyses of the material of the water spots revealed a marked high content of certain silicones, which clearly had exuded out of the coating surface and had been cured by irradiation during weathering. Overly clean water might support the formation of water spots.

However, the water should be pure enough to reliably exclude mineral deposits on the test panels or the inside of the test chamber. Reference values are provided in the respective weathering standards.

6.2.4 Simulation of acid atmospheric precipitation

As it became known that acid atmospheric precipitation caused the typical coating damage commonly referred to as acid etching on cars parked in the harbour area of Jacksonville, Florida, attempts were made to simulate this special kind of weathering stress in the laboratory, too. At first, traditional methods of testing chemical resistance in general were used to test resistance to acid atmospheric precipitation. A typical method that was introduced into testing of automotive coatings is the gradient oven method, which is designed to determine resistance to sulphuric acid. This method is still found in the test specifications of several automotive manufacturers.

For a better understanding of the weakness of this test, some background information about how it is performed is desirable. The test starts by depositing a fixed number of drops of diluted sulphuric acid (e.g. 1 %) at specified distances from each other (e.g. 1 cm) on the surface of a specially designed test panel. The acid drops are mostly deposited along the length of the test panel. The test panel is then exposed to thermal stress in a gradient oven for a fixed exposure period (e.g. 30 min). The gradient oven is designed such that each of the spots occupied by an acid drop is thermally stressed by a different temperature (e.g. between 30 °C and 80 °C). After exposure, the test panel is washed and inspected. The temperature of that spot, which is identified as being the last one to remain undamaged, serves as the criterion for acid resistance and is called the spot-free temperature. In other words, the higher the spot-free temperature, the greater is the acid resistance. Figure 6.34 compares the results of the gradient oven method with those of an outdoor weathering test carried out in Jacksonville, Florida, for the example of 12 black-coloured automotive coatings of different chemical composition.

The correlation between the two tests is poor, as shown by the small PEARSON coefficient of 0.6. Since it offers the advantage of low time consumption, many attempts have been made to improve the correlation of the gradient oven method with actual practice by varying the temperature, the exposure period, the composition of the acidic solution and the concentration of the acids. However, all these efforts have been unsuccessful.

As already mentioned in Chapter 3.2.6, acid atmospheric precipitation occurs not only in Jacksonville, Florida, but in all other industrial conurbations, too. Since 70 to 80 % of the rainfall can be classified as acidic, stress by acid atmospheric precipitation contributes substantially to coating ageing in normal use. Microscopic examination shows that the typical coating damage in Jacksonville, Florida, also occurs on test panels weathered outdoors in moderate climate as well, but to a much smaller and less striking extent. Nevertheless, acid atmospheric precipitation is found to have a marked influence on weatherability rankings during normal use in moderate climates. Acid atmospheric precipitation, especially in the form of dew and fog, acts near heavily trafficked streets in urban and industrial environments, because acid dew and fog are predominantly caused by automotive exhaust gases. Automobiles, in particular, suffer from acid dew and fog they

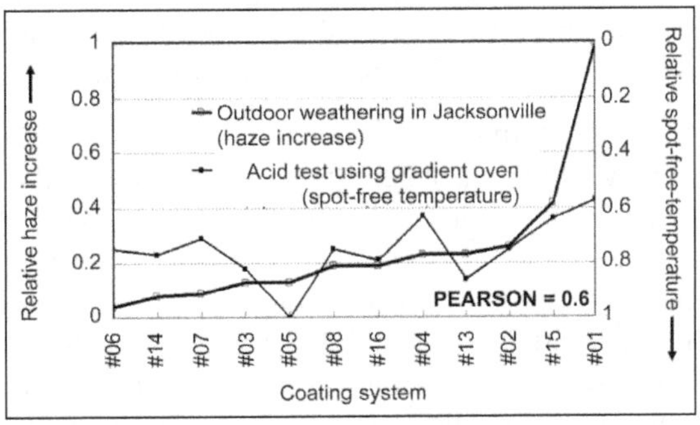

Figure 6.34: Comparison of the spot-free temperature determined in a gradient oven (30 min, 1 % sulphuric acid) as a common measure of acid resistance with the haze increase caused by 14-week conventional outdoor weathering in Jacksonville, Florida, for 12 different automotive coatings, adapted from [84]

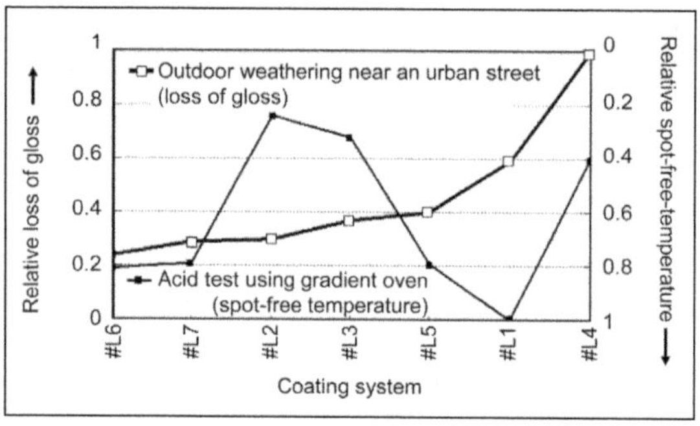

Figure 6.35: Comparison of the spot-free temperature determined in a gradient oven (30 min, 1 % sulphuric acid), which often serves as a measure of acid resistance, with the loss of gloss after 36 months of conventional outdoor weathering near a heavily trafficked urban street in Berlin, Germany, for 7 different types of clearcoat/ basecoat systems (coatings are not identical with those used for Figure 6.34), adapted from [85]

are parked for a long time on the roadside. Figure 6.35 compares the results of outdoor weathering near a heavy-trafficked urban street with those of the gradient oven method for seven different automotive coatings. Clearly, the coatings in the real environment did not behave like those in the gradient oven.

As shown by the different examples, the gradient oven method fails to allow correct predictions to be made about the weatherability of coatings intended for use in regions stressed by acid atmospheric precipitation. Furthermore, the highly subjective determination of the spot-free temperature is a major disadvantage. The main reason for the poor correlation between the gradient oven method and practice seems to be the fact that the coatings to be tested are only exposed to chemical and thermal stress. The synergy of acid atmospheric precipitation and global UV, as discussed in Chapter 3.2.6, is not included in the weathering stress simulation. Accordingly, it may be stated that **practice-oriented simulation of acid atmospheric precipitation can only take the form of a correspondingly extended or modified artificial weathering test**. Corresponding process developments subsequently focused, therefore, on the mutual action of UV and acid and were aimed at simulating all the weathering stress acting on automotive coatings in Jacksonville, Florida. Automotive coatings were chosen to be the subject of development because of their high significance in the context of this problem.

Three methods of artificial weathering which have reached significance in coatings testing will be described below:

- **ADF test** (Acid Dew and Fog test) of the Federal Institute of Material Testing and Research (BAM) Berlin, Germany, as per VDI Guideline 3958 Part 12 [86]
- **SPART 14** (SP Acid Rain Test facid=14) of the Swedish National Institute for Testing and Research (SP), Sweden [87]
- **BASF AAT** (Accelerated Acid Test) published by BASF (international) [88].

The design of the tests with regard to recommended test cycles and test parameters as well as recommended weathering devices was based on already standardized methods of artificial weathering. Details of the tests are described in Tables 6.8, 6.9 and 6.10 (page 148).

All three methods allow for the synergy of acid atmospheric precipitation and global UV by employing phases of simultaneous action of acid and UV during their test cycles, but as different percentages of the total exposure period. (BASF AAT: 50 %, ADF test: 42 %, SPART 14: 3 %). All reflect the special property of sulphuric acid, namely that it cannot evaporate at common environmental temperatures. However, the degree of the mutual action of acid and UV has a particular influence on the degree of damage and, therefore, the detectability and selectivity of the test, too. With regard to temperature, relative humidity, and time of wetness (TOW), the climatic conditions in Jacksonville, Florida, are considered to different extents. There are examples for all three tests which show good ranking correlation with outdoor weathering performed in Jacksonville. A clear differentiation cannot be made between the three tests with regard to their correlation with practice as the test results were not comparable due to the use of non-identical coating samples. The results of corresponding inter-laboratory comparisons and other kinds of proficiency-testing are not available yet.

Table 6.8: The 24-hour exposure cycle of the Acid Dew and Fog test (ADF test type J25), as per VDI 3958 Part 12 [85], for simulating weathering stress, including acid dew and fog, specifically simulates the subtropical climate in the harbour area of Jacksonville, Florida

ADF-test (Type J2.5) Test step		Duration	E_e-UV	Test climate
1	wetting by short spraying of acid solution (pH 2.5) onto the specimen surface (30...40 g solution per 1 m² exposed surface)	<1 min	dark	CT**: 35°C
2	dry test phase 1	9 h	(45 ± 5) W/m² at 300...400 nm, (fluorescent lamp)	BST* and CT**: 35 °C 75 % R.H.***
3	dry test phase 2	5 h	45 ± 5) W/m² at 300...400 nm, (fluorescent lamp)	BST* and CT**: 60 °C 40 % R.H.***
4	wet test phase (continuous wetting by neutral rain)	4 h	45 ± 5) W/m² at 300...400 nm, (fluorescent lamp)	CT**: 35 °C
5	dry test phase 3	6 h	(45 ± 5) W/m² at 300...400 nm, (fluorescent lamp)	BST* and CT**: 60 °C 40 % R.H.***

**Black standard temperature, **Chamber air temperature, ***Relative humidity of chamber atmosphere*

Simulated acid precipitation	acid dew and fog
Percentage of exposure period with acid	42 % (70 h per week)
Percentage of exposure period with acid + UV	42 % (70 h per week)
Percentage of exposure period with UV	100 % (164 h per week)
Recommended exposure period	28 cycles (20 with acid, 8 without acid)
Exposure position of the specimens	vertical
Assessment/ageing criterion	objective/gloss, haze, colour
Basic test standard	VDI guide 3958/ part 1 ;/ISO 4892-3,
Recommended weathering device	fluorescent UV lamp devices

Table 6.9: The 3-hour exposure cycle of the SP Acid Rain Test facid=14 (SPART 14) for simulating weathering stress, including acid rain, adapted from [86]

SPART 14 Test step		Duration	E_e-UV	Test climate
1	wet test phase 1(continuous wetting by acid rain (pH 3.2) at each 14th cycle by neutral rain at all other cycles)	20 min (on the front side)	0,5 W/m²/nm at 340 nm (xenon lamp)	CT**: 38 °C
2	dry test phase 1	60 min	0,5 W/m²/nm at 340 nm (xenon lamp)	BST*: 70 °C CT**: 47 °C 75 % R.H.***
3	wet test phase 2 (continuous wetting by neutral rain (all cycles))	60 min (on the back side)	dark	CT**: 38 °C 95 % R.H.***
4	dry test phase 2	40 min	0,5 W/m²/nm at 340 nm (xenon lamp)	BST*: 70 °C* CT**: 47 °C 75 % R.H.***

**Black standard temperature, **Chamber air temperature, ***Relative humidity of chamber atmosphere*

Simulated acid precipitation	acid rain
Percentage of exposure period with acid	3 %
Percentage of exposure period with acid + UV	3 %
Percentage of exposure period with UV	67 %
Recommended exposure period	till 25 % loss of gloss
Exposure position of the specimens	vertical
Assessment/ageing criterion	objective/gloss, colour
Basic test standard	SAE J 1960
Recommended weathering device	all weathering devices allowing wetting

Table 6.10: The 24 h-hour exposure cycle for the Accelerated Acid Test (BASF AAT) for simulating weathering stress, including acid rain, adapted from [87]

BASF AAT Test step		Duration	E_e-UV	Test climate
1	wet test phase 1 (continuous wetting by acid rain)	1 min	dark	
2	dry test phase 1	3 h + 50 min	dark	BPT* and CT**: 38 °C 80 % R.H.***
3	dry test phase 2	12 h	0,55 W/m²/nm at 340 nm, (xenon lamp)	BPT*: 80 °C CT**: 55 °C 80 % R.H.***
4	dry test phase 3	27 min	dark	BPT* and CT**: 38 °C 80 % R.H.***
5	wet test phase 2 (continuous wetting by neutral rain)	1 min	dark	
6	dry test phase 4	3 h + 50 min	dark	BPT* and CT**: 38 °C 80 % R.H.***
7	wet test phase 3 (continuous wetting by neutral rain)	1 min	dark	
8	dry test phase 5	3 h + 50 min	dark	BPT* and CT**: 38 °C 80 % R.H.***

**Black panel temperature, **Chamber air temperature, ***Relative humidity of chamber atmosphere*

Simulated acid precipitation	acid rain
Percentage of exposure period with acid	68 %
Percentage of exposure period with acid + UV	50 %
Percentage of exposure period with UV	50 %
Recommended exposure period	till 700 h (valuation after each 100th h)
Exposure position of the specimens	horizontal
Assessment/ageing criterion	subjective/visual assessment
Basic test standard	ISO 4892-2
Recommended weathering device	xenon lamp devices

In view of the advanced state of national and international standardisation and the particular significance of acidic dew and fog in the occurrence of acid etching, only the ADF test will be discussed in more detail. This will create a better understanding of the problems arising with simulating the action of acid atmospheric precipitation by artificial weathering. As in the case of every other arbitrary method of artificial weathering, the ADF test was designed to reflect the fact that usable results can be achieved faster in the laboratory than outdoors only if the test is performed in time-compressed or accelerated fashion.

The general method employed in environmental testing is that of test tailoring, described in VDI Guideline 3958 Part 1 [9] and discussed in Chapter 2.5. Recall that the guideline recommends effecting time-compression by applying the worst case weather scenario or scenarios for the coatings to be tested at a much higher frequency than occurs in normal use, but without overly exceeding the testing parameters! Knowledge of the applicable worst case weather scenarios is a prerequisite for correctly designing the weathering cycle of the laboratory test. As described in more detail in Chapter 3.2.6, the mutual action of global UV and acid atmospheric precipitation proves to be one of the key worst case weather scenarios for modern coating systems for use in acid-stressed environments (see Table 2.4). Thus, the goal in designing the ADF test was to combine the aforementioned worst case weather scenario with the other weathering stresses, such as heat and water, in a reasonable manner and to tailor them to automotive coatings.

The recommended composition of the simulated acid dew and fog is a mixture of sulphuric, nitric and hydrochloric acids (mass ratio $H_2SO_4 : HNO_3 : HCl = 1 : 03 : 0.17$) diluted to a pH value of 2.5. Recall that the same solution is used for the Southern Florida Acid Rain Test, described in Chapter 4.3.7. A pH value of 2.5 appears to be optimal for artificial weathering, too. A pH of 2.5 corresponds to the maximum acid concentration regularly measured in the rain clouds over Mt. Brocken.

Let us now discuss the **24-hour ADF** cycle in Table 6.8. The first step is wetting of the test panels by brief spraying of the artificial acid dew and fog in order to simulate the acid atmospheric precipitation occurring in the harbour area of Jacksonville, Florida. Dew and fog having 10-times higher acid concentration on average than acid rain are simulated because they are the most aggressive and most frequent form of wetting in natural environmental conditions. Therefore, one of the essential differences from the other two acid test methods is the way in which acid atmospheric precipitation is simulated. In contrast to the ADF test, the other two methods claim to mostly simulate the action of acid rain.

As shown in Figure 6.36, the chosen manner of applying the simulated dew by brief spraying of the test panel surface once every ADF cycle leads to the same surface density and size of drops daily observed on car roofs naturally wetted by dew or fog in the early morning hours.

Again due to the form in which the acid solution is applied, the ADF test leads to almost homogeneously distributed acid etching, as shown in Figure 6.37 for two different automotive coatings (#05 and #16). The coatings' different weatherability, as revealed by the different density of the acid etching, corresponds to the values of haze increase also shown in the diagram representation. The good homogeneity of the coating damage permits the use of objective methods of measurement, such as determinations of haze and gloss. Objective assessment of the coating damage is a further important advantage of the ADF test. The other two methods only use highly subjective visual assessment. The argument often cited against the ADF test that the damage deviates markedly from that found after outdoor weathering in Jacksonville, Florida, is not valid with regard to modern automotive coatings any more. A study performed on 20 different basecoat/clearcoat systems for automotive OEMs produced in 1999 revealed that, after conventional outdoor weathering in Jacksonville, Florida, only roughly half the coatings displayed the typical kind of damage shown in Figure 3.52 while the other half almost all displayed acid etching identical in shape and size with that which occurred during the ADF test [84]. More or less large craters, distributed almost homogeneously across the test panel surface after the Jacksonville exposure and often adduced as an argument against the ADF test, can be found after both natural and artificial weathering.

Figure 6.36: Surface of a Chapter 5 cm in length of a test panel after wetting by the defined short spraying of the acid solution (test step 1 of the ADF test) with a spray gun similar to a common perfume atomiser

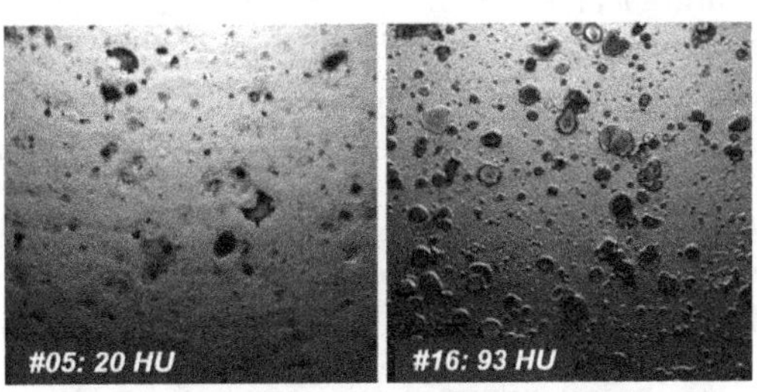

Figure 6.37: Surfaces of 14-mm chapters of two automotive coatings of different acid resistance (#05 and #16) showing typical acid etching damage caused by 28 days of ADF test, along with the haze increase (in Haze units HU)

Now, let us discuss the further steps of the ADF cycle. The two dry phases (steps 2 and 3) following the spraying phase (step 1) are intended to simulate the synergy of acid and UV as representing the worst case weather scenario mostly

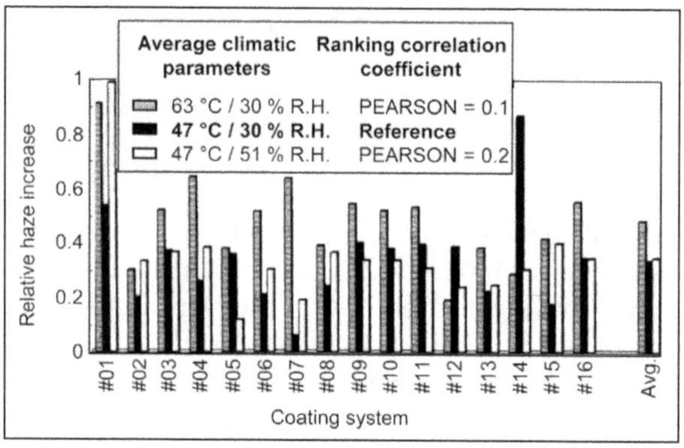

Figure 6.38: Influence of temperature and relative humidity on the acid etch damage caused by the ADF test in a GUT200 fluorescent lamp weathering device presented as the increase in haze of 16 differently acid resistant black-coloured automotive coatings

occurring during the summer season in Florida. The data for the test climate in the last column pertain to the coating surface micro climate. The two chosen different combinations of temperature and relative humidity characterise two weather scenarios that are typical of Florida and are empirically known to dominate the rate of coating ageing. These are the most frequently changing weather scenarios outdoors that involve stress by direct and diffuse irradiation. Step 3 (dry phase 2), comprising 5 hours at 60 °C and 40 % R.H., simulates the natural phases of stress by direct irradiation when the surface temperature reaches its highest and the relative humidity on the coating surface its lowest value. Step 2 (dry phase 1), comprising 9 hours at 35 °C and 75 % R.H., simulates the natural phases of stress by diffuse irradiation when the surface temperature and the relative humidity on the coating surface reach almost ambient values. As already described in Chapter 3.2.5, a weather scenario comprising stress by diffuse radiation only occurs, for instance, when the sun is covered by a cloud in a sky with 50 % cloud cover. Simulation of both weather situations is necessary because the coating surface micro climate influences the coating ageing to different extents as a function of the coating type. The extent to which temperature and relative humidity influence the results of the ADF test is shown in Figure 6.38. Identical test series consisting of 16 black-coloured basecoat/clearcoat systems for automotive OEM were exposed to three test variants differing in the chosen combination of temperature and relative humidity. The average values of the two climatic test parameters over the whole ADF cycle were used. Haze increase served as ageing criterion, and is presented by the coating numbers.

In this diagram, the result of a central test variant, indicated by the black bars, is compared with the result of a further test variant (shaded bars) performed at a 16 °C higher temperature (but identical relative humidity of 30 %) and with the result of a further test variant (white bars) performed at a 21 % higher relative humidity (but identical temperature of 47 °C). The chart clearly shows that an increase in one of the two climatic test parameters can boost as well as weaken the synergy of UV and acid as a function of the type of the coating to be tested. About half of the coatings involved in the test series responded with higher damage if temperature or relative humidity was increased relative to the central test variant, for instance, coating #01. In contrast, other coatings, such as coating #14, exhibited the reverse behaviour, for at least one of the climatic test parameters. Comparison of these results readily reveals that the rankings of the coatings for expected damage will change enormously if the temperature or relative humidity varies. The PEARSON coefficient, worse than 0.2, proves that the chosen three test variants did not correlate to each other.

Recall that the marked influence of the coating surface micro climate on the degree of damage is also one of the most essential reasons for the poor repeatability of the outdoor weathering in Jacksonville, Florida, as mentioned in Chapter 4.2.3.

The subsequent wet phase (step 4 of the ADF test), characterised by continuously applied neutral rain (demineralised water), interferes with the mutual action of acid and UV. The duration of raining is practice-oriented and considers possible cyclical mechanical stress caused in the coating film by changes between water uptake and drying. This test step also allows for certain coating additives, such as light stabilisers, to be transferred in water-soluble form due to acid attack and

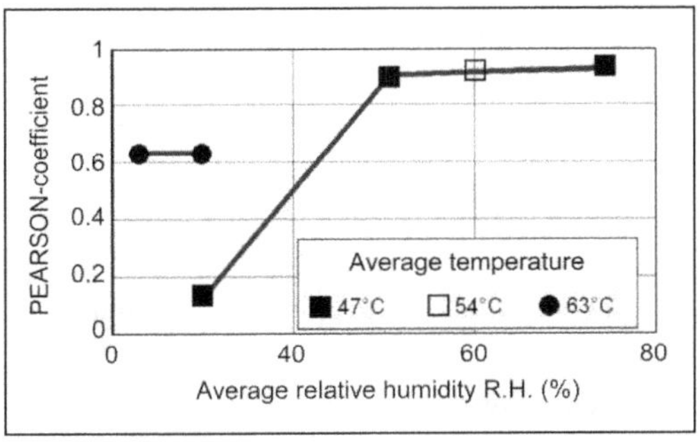

Figure 6.39: Influence of the test climate in the ADF test on the correlation with outdoor weathering in Jacksonville, Florida, presented in the form of a PEARSON ranking correlation coefficient as a function of the average values of temperature and relative humidity in which increase in haze acts as ageing criterion

then removed by the water. Loss of light stabilisers can lead to loss of protection against UV and faster coating ageing from cycle to cycle. Regular interruption of acid attack by the rain phase is also necessary to approximate to the ratio of weather phases with and without acid stress, as found in reality. There is an optimal ratio, compliance with which is necessary for good correlation with practice.

Step 5, the third dry phase to be performed acid-free, serves the same purpose. Additionally, five ADF cycles are always followed by two identical test cycles conducted acid-free by excluding step 1. This is also thought not to unrealistically exceed acid stress and to be a suitable approximation for a normal week.

In general, dry phases during the ADF test reflect the experience that acid-damaged surface areas suffer more from UV than those not stressed by the acid drops, obviously as a result of acid-induced deterioration of light stabilisers. In contrast, undamaged surface areas mostly surrounding the acid etchings often experience a considerable increase in their acid resistance. How strongly such effects influence the test results depend on the coating to be tested.

The strong influence of the coating surface micro climate on the correlation between the test result and practice, as shown in Figure 6.38, requires precise definition of the tolerances for the test parameters that must be complied with. Figure 6.39 shows the PEARSON coefficients attainable with the ADF test in comparison to outdoor weathering in Jacksonville, Florida. They are presented here as a function of the test climate for six ADF test variants differing in temperature and relative humidity. The presentation is based on the same coating series used for Figure 6.38.

In order to reach good correlation with the Jacksonville outdoor weathering, the average surface temperature should be kept between 47 °C and 54 °C and the average relative humidity on the coating surface must not be lower than 50 %. The letter "J" in the acronym of ADF 3958 Part 12 [86] means suitable for simulating the weathering stress acting in Jacksonville, Florida. If these tolerances are observed, the ADF test can be performed with the same good correlation with practice in each type of weathering device, even in xenon lamp devices [80]. Figure 6.40 shows

ADF test (method J2.5)	#13	#05	#07	#12	#14	#06	#08	#03	#09	#11	#10	#16	#02	#04	#15	#01
Rank	1	2	3	4	5	6	7	8	9	10	11	12	13	14	15	16
Jacksonville (outdoors)	#06	#14	#07	#03	#05	#11	#12	#16	#08	#10	#04	#09	#13	#02	#15	#01

Weatherability: #XX good | #XX medium | **#XX** bad

Figure 6.40: Simplified representation of the attainable ranking correlation between Jacksonville outdoor weathering and ADF Test J25 for 16 coating systems (#01 to #16) arranged in decreasing order of weatherability for both weathering tests, and classified as "good", "medium" and "poor"

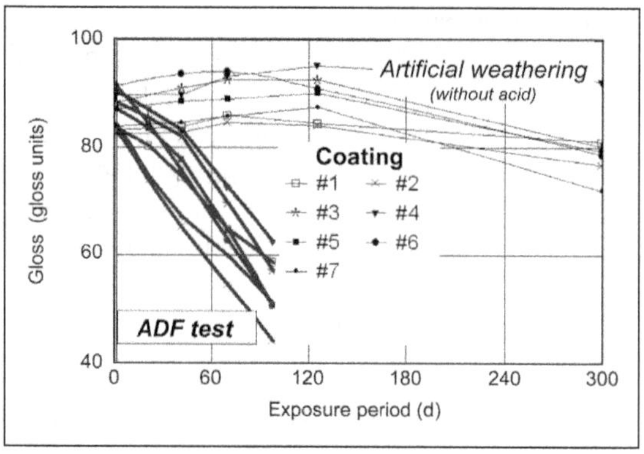

Figure 6.41: Accelerating effect of acid stress in artificial weathering shown by comparing the results of the ADF test with those of an acid-free weathering test performed with the same climatic test parameters for a series of 7 different basecoat/clearcoat systems for automotive OEM in which gloss served as ageing criterion, from [90]

that automotive coatings can be graded by means of the ADF test into the following three weatherability groups: good, medium and poor.

In line with this presentation, it can be stated that only one (#13) out of the sixteen coatings involved in the studies was falsely assessed on the basis of the ADF test concerning the weatherability expected in Jacksonville, Florida.

As shown on Mt. Brocken, most rainfall can be graded as acidic, not only in Jacksonville, Florida, but also in other areas of the globe. That might be the reason for the good correlation between the ADF test and outdoor weathering, even when carried out in less acid-stressed sites. For instance, the ADF test is successfully used in the testing of aircraft coatings [52]. Due to the high influence of the coating surface micro climate on the weathering results, a further ADF test variant (M2.5 see VDI Guideline 3958 Part 12 [86]) is recommended for coatings used on aircraft or in regions with moderate climate. The moderate variant (M2.5) differs from the Jacksonville variant (J2.5) only in having lower relative humidity.

The accelerating effect caused by including acid stress in the artificial weathering is demonstrated in Figure 6.40 for seven automotive coatings that vary substantially in type and application. It shows the gloss values during the ADF test (thick lines) and a further artificial weathering test that works without acid stress (thin lines). The two tests were performed with identical climatic test parameters.

Figure 6.41 shows a ten-fold acceleration in coating ageing, simply due to the inclusion of acid stress in the artificial weathering test. The weatherability ranking of the seven coatings was not influenced by the presence of acid. In other words, **including acid stress in the artificial weathering permits not only practice-oriented simulation of the special weathering stress of Jacksonville, Florida, but also substantial acceleration, relative to the common methods of artificial weathering, without loss of correlation. The ADF test therefore is a typical form of accelerated testing.**

6.3 Acceleration and correlation of artificial weathering with practice

Typical methods of artificial weathering are shown in Tables 6.1 and 6.2. As can be seen, they are aggravated relative to the natural weathering stress in different ways and to different extents. Each of the test method claims to best simulate the natural weathering stress occurring in normal use and to be optimal for accelerating coating ageing without any loss of correlation with practice. Naturally, all of the test methods listed in Tables 6.1 and 6.2 also claim to attain the specified limits of the corresponding ageing criteria (failure criteria) considerably faster than in normal use. However, with present testing practice, coatings that prove to be one of the best in artificial weathering often prove to be one of the worst in normal use and vice versa. It has also been observed that the same coatings tested by a further artificial weathering method yield good correlation with practice by chance. When the applicable ageing criteria change too slowly, as is also often observed, the artificial weathering method employed may not have created the worst case weather scenario applicable for the coatings to be tested. On the other hand, adequately accelerated test methods often show poor correlation with practice.

According to the current state of knowledge, the reason that weathering results do not agree with those of normal use can be seen in degradation processes, mostly those determining ageing behaviour, which do not respond in the same measure to the degree to which the weathering stresses are aggravated. Only in rare cases are linearly functional correlations found between cause and response in coating ageing. The discussions in Chapter 6.2.2.4 und 6.2.2.5 confirm this statement for photon energy and irradiance. Furthermore, different ageing criteria, such as loss of gloss or change in colour, can respond to different extents and in different ways to exaggeration of the weathering stresses. It is generally true that **the more accelerated the artificial weathering – the worse is its correlation with practice.**

The optimum exaggeration of weathering stress is always a compromise between acceleration and correlation. Corresponding experience has shown that a 5-fold acceleration relative to outdoor weathering in southern Florida and a 10-fold acceleration relative to outdoor weathering in central Europe provide acceptable correlation with practice. Furthermore, indefinable relationships between ageing effect and exposure period often need to be considered, as well. Therefore, the use of weathering results for predicting service life can be complicated even further. First, let us discuss the problems resulting from differently overly exceeded testing parameters. Figure 6.42 attempts to illustrate the problems in a simplified, understandable manner.

Two coating types denoted by A and B (e.g. TSA and 2-pack PUR) were chosen as typical examples. They are characterised by non-linear and variable correlation between the factor aggravation (f_i) of weathering stress and the resulting acceleration of coating ageing. Two different ageing criteria denoted by 1 and 2 (e.g. loss of gloss and change in colour) are to be used to assess the acceleration. The chart on the left hand side demonstrates the influence of the coating type for ageing criterion 1, e.g. loss of gloss. A and B should differ markedly in the slope of their curves. This means that A and B respond to different extents to increasing artificial weathering stress (e.g. by increasing irradiance photon energy or temperature) relative to normal use. For simplicity,

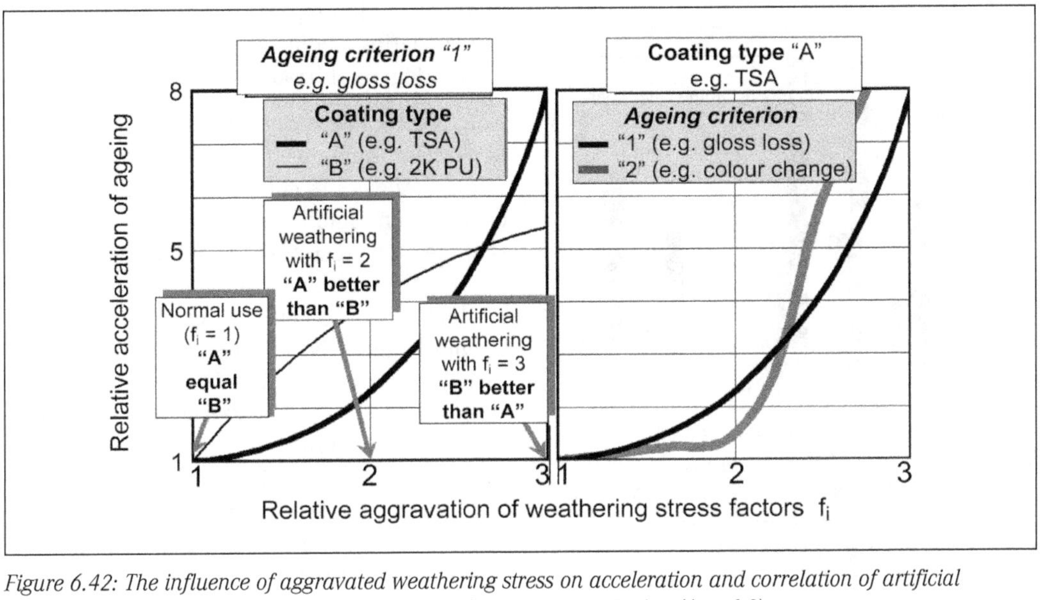

Figure 6.42: The influence of aggravated weathering stress on acceleration and correlation of artificial weathering as a function of coating type (A and B) and ageing criterion (1 and 2)

assume that A and B do not differ in their ageing behaviour in normal use, i.e. both curves start at the same point (1/1). Despite identical ageing behaviour in normal use, the curves begin to drift apart and then cross with increasing aggravation factor. Now look at how the different shape of the curves influences the behaviour of the two coating types. Consider first the situation for two-fold aggravation. In this situation, the ageing of coating A is accelerated much less than that of coating B. This creates the impression that coating A has the better weatherability. Now, consider the situation for three-fold aggravation. In this situation, ageing of coating B accelerated much less than coating A. This creates the impression that coating B is better. Only at roughly 2.5-fold aggravation, where both coating types signal the same acceleration, does artificial weathering correlate with normal use.

The right hand diagram demonstrates the influence of the chosen ageing criterion for coating A. It is also known that different types of photochemical reactions are responsible for the ageing-induced changes of different coating properties, such as gloss and colour. Different photochemical reactions are mostly influenced to different extents by weathering stress. Therefore, different ageing criteria can be expected to respond to different extents to increasing testing parameters. As demonstrated, the curves for the two ageing criteria deviate markedly from each other as well as crossing each other. Thus, loss of gloss is influenced more extensively than change in colour for two-fold aggravation. However, for three-fold aggravation, the reverse behaviour is observed for the two ageing criteria.

In other words, **the ranking of coatings tested together during artificial weathering depends not only on the degree of aggravation but also on the chosen ageing criterion**. Since the test methods currently used in artificial weathering often employ unnaturally exceeded photon energy, different degrees of irradiance and different coating surface micro climate, considerable differences in the weatherability rankings must be expected for identical series of coatings exposed to different test methods. This applies especially to coatings that differ in resin and additives.

Now, let us look at some real examples in coatings testing. Figure 6.43 shows for the example of two clearcoat samples that differ only in the kind of stabilisation how strongly the weathering method determines the weathering results. The ageing criterion was the change in colour. As expected, methods F_{2a}, F_{2b} and X_{1a} (see Table 6.2 for description) exhibited the fastest coating ageing, obviously because of the presence of UV radiation below 300 nm. Comparison of the methods F_{2a} and F_{2b} method F_{2a} shows that the faster coating ageing is obviously due to the higher degree of irradiance.

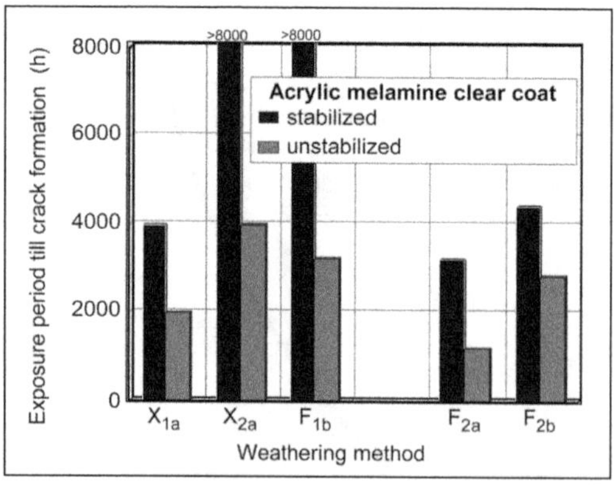

Figure 6.43: The influence of the weathering test method on the ageing behaviour of two differently stabilized samples of a waterborne acrylic melamine clearcoat applied to water-borne basecoat, presented as exposure period up to crack formation. The weathering test methods employed are explained in Table 6.1, adapted from [16]

As expected, the presence of light stabilisers led to slower coating ageing in all artificial weathering methods. As can also be seen, the presence of light stabilisers did not influence the ranking of the results of the different test methods. In other words, for both clearcoat samples, method X_{2a} is the most benign and method F_{2a} the most aggressive one. Figure 6.44 shows the results obtained by a different laboratory for a further example [22]. Two 2-pack PUR clearcoats differing in the types of polyol (A and B) were used in the study. Each of the clearcoat types was stabilised in a different way (see the legend to Figure 6.44). The exposure period up to identical yellowing ($\Delta b^* = 2$) is shown as a function of the weathering method.

Just as in the aforementioned example, the weathering methods differ in their aggressiveness. The influence of the cut-on of the used radiation is clear. Where the clearcoat samples differ only in the type of stabilisation, the different weathering methods reveal sufficient correlation with each other. Figure 6.45 presents the results of the same studies in a different way. The results of xenon lamp device weatherings are compared to the results of outdoor weathering carried out in southern Florida. The exposure period needed for yellowing of $\Delta b^* = 2$ served as ageing criterion.

Also in this presentation, method X_{1a} yields the fastest yellowing for the same reason (cut-on <300 nm). Comparison of the artificial weathering methods shows that the slowest yellowing in method, X_{2b}, might be caused by the smallest TOW or the lowest coating surface temperature (BST), as can be seen in Table 6.2. The slower yellowing in outdoor weathering, in spite of identical radiant exposure, indicates enhancement of other weathering stresses in artificial weathering, too. A look at the ranking of the four clearcoat samples reveals that the three artificial weathering methods correlate quite well with the outdoor weathering in this case.

The use of the results of artificial weathering to predict ageing behaviour in normal use is a second major problem of coatings testing in general. The reason for difficulties in transferring the results of artificial weathering to practice also lie in the functionally undefined correlation between coating ageing and radiant exposure. As discussed, before this correlation depends on the type of coating to be tested as well as the ageing criterion chosen for assessing the ageing behaviour. Figure 6.46 (page 156) shows the different changes in yellowing and loss of gloss measured during method X_{1a} for the example of clearcoat P-0. In this case, loss of gloss was noticeable only after 1000 hours of exposure. In contrast, marked yellowing occurred during the first 100 hours of exposure and reached saturation after about 1200 hours. **The lack of agreement between the curves of weathering-induced changes in colour and gloss are typical of automotive coatings testing.**

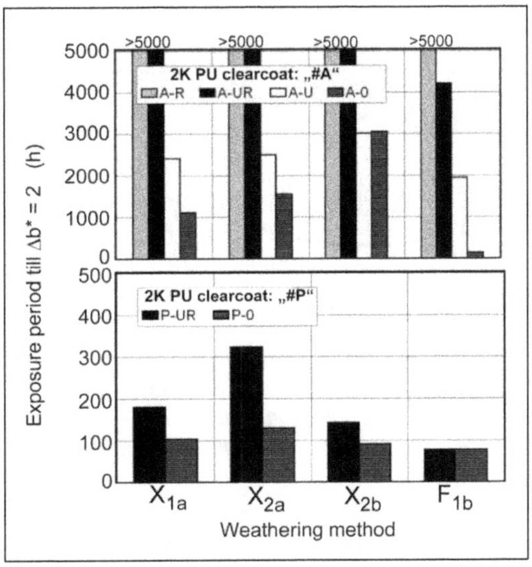

Figure 6.44: Influence of weathering test method on the ageing behaviour of six 2-pack PUR clearcoats differing in polyol (A in the top chart and P in the bottom chart), as well as in the type of stabilisation (0: unstabilised, U: containing UV absorbers only, R: containing only radical scavengers, UR: containing UV absorbers and free-radical scavengers) in which exposure period up to yellowing $\Delta b^* = 2$ serves as ageing criterion. The weathering test methods employed are described in Table 6.2, adapted from [22]

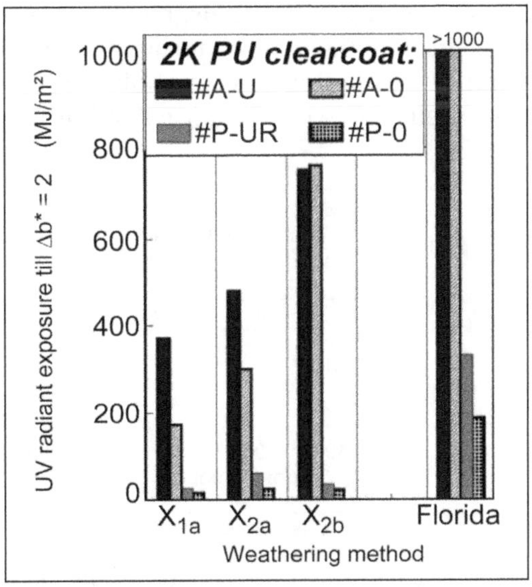

Figure 6.45: Influence of the weathering test method on the ageing behaviour of four 2-pack PUR clearcoats differing in the polyol (A and P) as well as in the type of stabilisation (0: unstabilised, U: containing only UV absorbers, "UR": containing UV absorbers and free-radical scavengers) in which the exposure period up to yellowing $\Delta b^* = 2$ serves as ageing criterion The artificial weathering methods employed are explained in Table 6.2, adapted from [22]

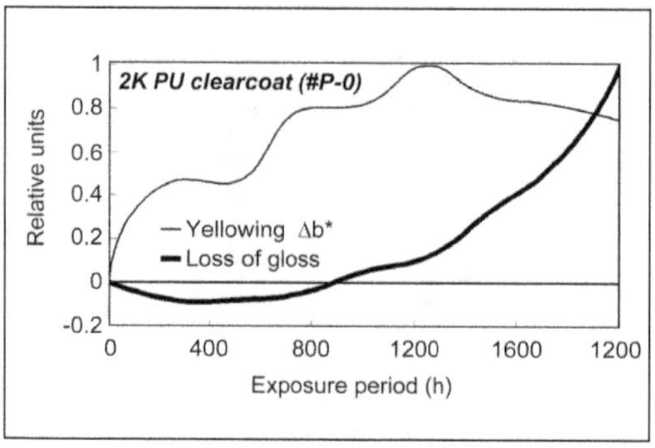

Figure 6.46: Loss of gloss and yellowing of an unstabilised 2-pack PUR clearcoat (sample #P-0) during the weathering test method X_{1a} explained in Table 6.2, adapted from [22]

Figure 6.47 compares artificial and outdoor weathering for the case of weathering-induced yellowing and loss of gloss as a function of exposure period for the same clearcoat sample (P-0). It confirms differently shaped curves for different ageing criteria, as already partly shown in Figure 6.46.

Consider first the ageing criterion yellowing. It has an almost linear shape and reaches $\Delta b^* = 7$ after 27 months of outdoor exposure (bottom left chart). If we look for the exposure period necessary for attaining the same value of yellowing in artificial weathering, all four methods in this presentation, probably by accident, reveal almost the same test length of 1000 hours (see top left chart). This computes to an acceleration factor of about 20 for artificial weathering relative to Florida outdoor weathering. However, as may also be seen, if other values of yellowing greater or smaller than $\Delta b^* = 7$ are specified as the failure criterion, greater or smaller acceleration factors different from that in the other artificial weathering method could result. The charts on the right of Figure 6.47 illustrate the relationship for loss of gloss as ageing criterion. In contrast to yellowing, the curve for the gloss loss during outdoor weathering deviates markedly from linearity. The highest attainable acceleration factor of about 10 was found at 2000 hours' exposure with method X_{1a}. However, if a 25 % loss of gloss is specified as the failure criterion, as common in coatings testing, considerably smaller acceleration factors are attainable, e.g. around 4 for method X_{1a} and 3 for method F_{2b}. The chosen examples show that, **due to the non-linearity between ageing criterion, degree of aggravation, and exposure period, and because of the unpredictable fluctuation in coating-related ageing processes, the acceleration factors in artificial weathering can vary substantially as a function of weathering method, coating type, ageing criterion and specified limit of acceptance (failure criterion).**

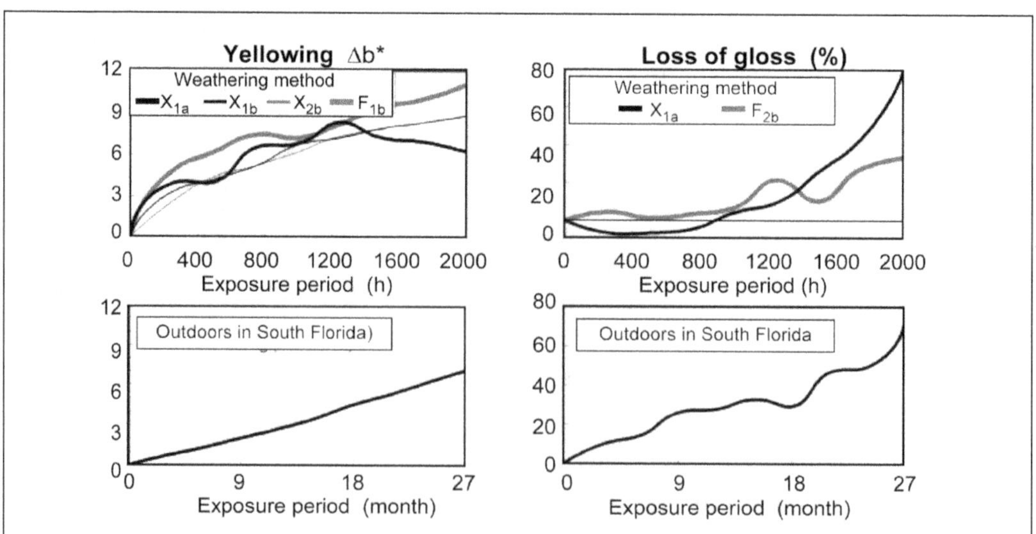

Figure 6.47: Yellowing and loss of gloss by an unstabilised 2-pack PU clearcoat (sample P-0) during different artificial weathering methods, as well as conventional outdoor weathering in southern Florida. The artificial weathering methods employed are explained in Table 6.2, adapted from [22]

Fewer problems are seen in monitoring the quality of series production where specimens of identical coatings are always exposed to artificial weathering, ranking correlation plays a subordinate role to the acceleration factors.

For judicious selection of appropriate weathering device, weathering method, exposure period, and ageing criteria, preliminary studies, usually outdoor weathering tests, are required. The exposure period should be long enough to reach the specified failure criterion, such as 25 % loss of gloss, start of crack formation or $\Delta E^*=2$. Calculation of the acceleration factor based on length of exposure to failure permits the coating's service life to be predicted with the lowest degree of uncertainty. However, it is also common to specify the length of exposure up to the point at which the specified ageing criterion just fails to be met, e.g. formation of cracks, residual gloss <75 %, or change in colour $\Delta E^*<2$. If the artificial weathering is monitored by reference materials as described in Chapter 6.2.2.8, there can be no objections to this approach.

6.4 Devices for artificial weathering

In line with Chapter 6.2.2.1, the following discussions are restricted only to weathering devices employing xenon or fluorescent UV lamps. These are the most commonly employed devices in coatings testing. Different design concepts are used for weathering devices. They differ mostly in the positioning of the lamp and test panel inside the test chamber. The three most frequently concepts are shown in Figure 6.48.

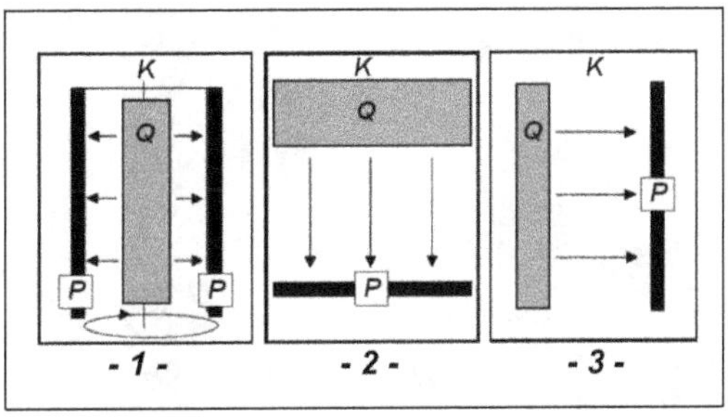

Figure 6.48: Most frequently employed concepts for positioning the radiation source (Q) and test panel (P) inside the test chamber (K) of weathering devices: 1: Lamp and test panel are both arranged vertically; the test panels rotate around the lamp. 2: Lamp and test panel are both arranged horizontally (stationary). 3: Lamp and test panel are both arranged horizontally (stationary)

Of course, the test panel faces the lamp in all weathering devices. Concerning their position in the test chamber, the three design concepts agree in the position of both xenon lamp and fluorescent UV lamp devices. Concept 1, characterised by vertical radiation source and rotation of the test panels around the lamp, is only used in xenon lamp devices. The Ci5000 Xenon "Weather-o-meter" from ATLAS Electric Dev. Co (USA) shown in Figure 6.49 is a typical example of concept 1.

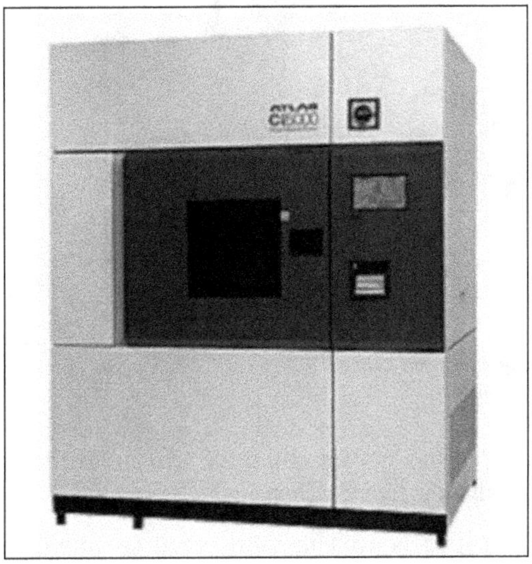

Figure 6.49: The Ci5000 Xenon "Weather-O-meter" is a typical example of design concept 1,(from [30])

Figure 6.50 (page 158) shows a look inside the test chamber of the Ci5000. A 12000 W water-cooled xenon arc lamp system (X) is arranged in the centre of a nearly spherical basket-like specimen rack. The water-cooled xenon lamp is surrounded by two cylindrical optical filters. Cooling water flows in between the xenon lamp, and an inner filter and an outer filter. In addition to its cooling function, the water absorbs some of the unwanted near infrared radiation (NIR). More NIR can be removed with special infrared absorbing or reflecting filters.

Figure 6.50: Positioning of xenon lamp (X) and test panels (P) in the Ci5000 Xenon "Weather-o-meter", from [30]

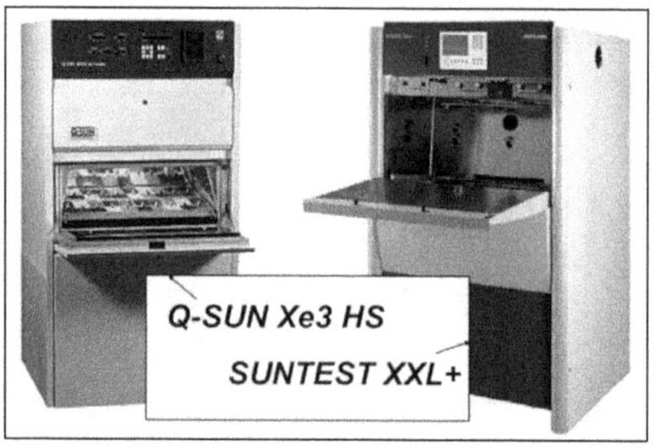

Figure 6.51: Two modern xenon lamp devices typical of design concept 2, adapted from [30] and [91]

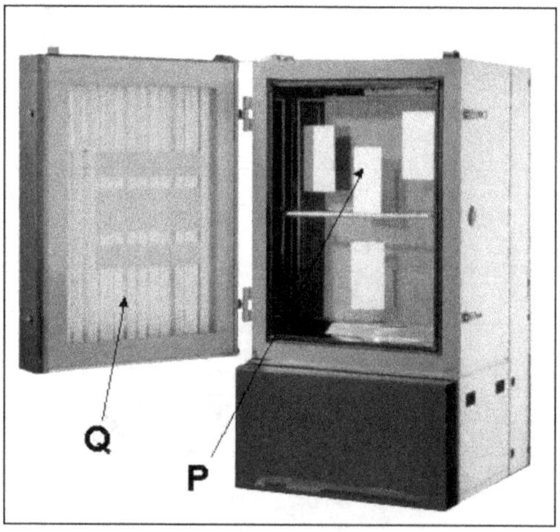

Figure 6.52: Global UV Test Unit, model UV200, from [75]

The specimen rack rotates around the xenon lamp system. The test panels are arranged on three tiers inside the specimen rack. Through this type of positioning, only the test panels on the middle of the three tiers are exactly vertical when facing the xenon lamp. The test panels on the other two tiers face the lamp at an angle. During irradiation, the test panels and the specimen rack rotate around the xenon lamp system continuously at low speed. The total exposure area is 1.1 m^2. Irradiance can optionally be regulated to twice the value of the global UV, based on narrowband (340 nm or 420 nm) or broadband (300 to 400 nm) values determined at specimen level. Black panel temperature or black standard temperature (both on the specimen rack) can automatically be controlled in the range from 40 °C to 110 °C (BPT) or 40 °C to 120 °C (BST) by means of automatic blower speed control. Further equipment serves to regulate the climate inside the test chamber, e.g. the "VibraSonic" Humidification System and special air heaters. The corresponding climatic working and controlling range is shown in Figure 6.10. Furthermore, the Ci5000 is equipped with a spray system for wetting the test panels with water during irradiation. The lamp has to be replaced after 1200 hours of burning. The xenon lamp devices "SUNTEST XXL"+ from ATLAS Electric Dev. Co (USA) and Q-SUN Xe3 HS from Q-Panel Co (USA) are good examples of design concept 2, employing horizontal radiation systems. These are shown in Figure 6.51.

Both devices work with three air-cooled, 2KW, xenon lamps in combination with optical filter systems. Due to the horizontal position of the lamps and immobilizing of the test panels, concept 2 does not need the complicated and mostly vulnerable mechanical rotation system. The exposure area, which can mostly be tilted up to ±5°permits the weathering stress of slightly angled coated areas in normal use, such as automobile roofs, to be simulated. Both devices have a total exposure area of about 0.3 m^2. Equipped with means of controlling the climate in the chamber and wetting the test panels, both are also highly effective weathering devices.

Concept 3 is predominantly used for fluorescent UV lamp devices. Global UV Test Unit model UV200 from WEISS Umwelttechnik GmbH (Germany) QUV/spray from Q-Panel Co (USA) and UV2000 from ATLAS Electric Dev. Co (USA) are typical examples.

Figure 6.52 shows the Global UV Test Unit, model UV200 fluorescent UV lamp device (GUT200 in short). As can be seen through the opened chamber door, the fluorescent UV lamps (Q) and the test panels (P) are both vertical. Eighteen fluorescent UV lamps are housed in the chamber door to create an almost two-dimensional radiation source. The weathering device can be used to perform all the weathering methods described in ISO 4892 Part 3 whether equipped with identical lamps, e.g. UV-A-340 or with a UV lamp combination of four types of lamps with differently positioned irradiance maxima (313, 340, 365 and 420 nm). Radiation source and test chamber are separated by a pane of material which transmits the UV radiation in accordance with ISO 4892 Part 3. A total exposure area of about 0.75 m^2 is available. UV irradiance (300 to 400 nm) is 40 to 70 W/m^2. The radiation spectra resulting from the use of different fluorescent UV lamps can be seen in Figure 6.3. The climatic working and controlling range of the GUT200 is shown in Figure 6.11. The two other fluorescent UV lamp devices mentioned earlier are shown in Figure 6.53.

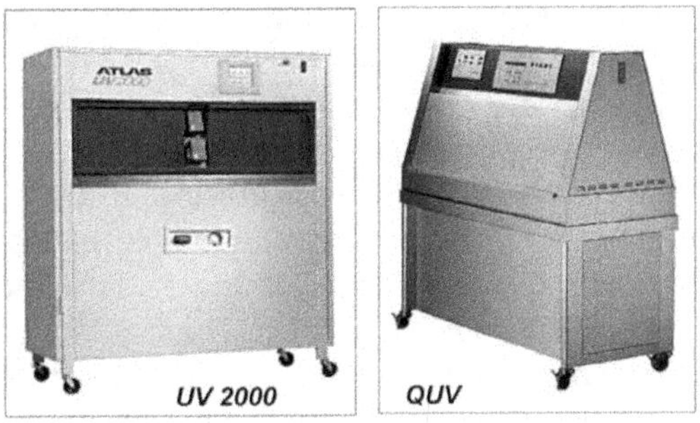

Figure 6.53: UV2000 and QUV Fluorescent UV lamp devices, from [30] and [91]

The sketch in Figure 6.54 for both devices shows clearly that they belong to design concept 3.

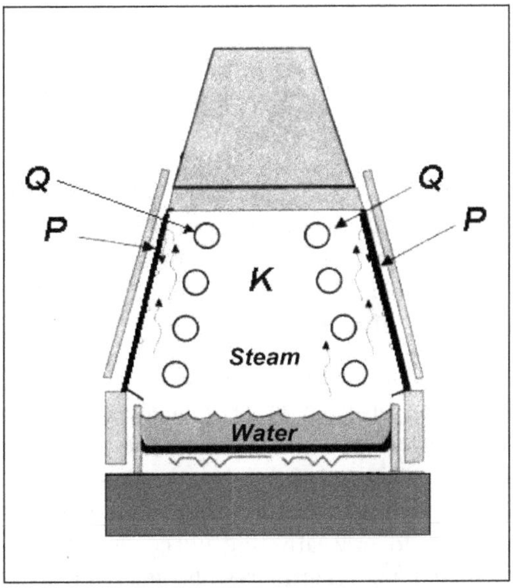

Figure 6.54: Positioning of lamps and test panels in both the UV2000 and QUV fluorescent UV lamp devices shown in Figure 6.53 (Q: fluorescent UV lamps, P: Test panels, K: Test chamber), adapted from [74] and [91]

Both weathering devices are equipped with two rows of eight 40W-fluorescent UV lamps (Q), as shown in Figure 6.54. The backs of the exposed test panels (P) can be cooled. In conjunction with high relative humidity inside the test chamber (K), wetting on the coating surface by condensation can only be attained by test panels with good thermal conductivity. To also allow wetting by artificial rain, both devices are additionally equipped with corresponding spraying systems. The most important standardized methods of artificial weathering as well as basic studies of coatings' ageing behaviour can be generally carried out in the aforementioned weathering devices. However, beware of weathering devices of overly limited performance, as false predictions about the coatings' service life may result!

6.5 Advantages and disadvantages of artificial weathering

The enormous shortening of the time needed to gain information for predicting the service life of the coatings to be tested can be seen as the greatest advantage of artificial weathering over outdoor weathering. This applies not only to the exposure period but also to the fact that the **start of exposure is not restricted** to a specific season. For instance, outdoor weathering in Jacksonville, Florida, requires that the test panels always be available at the weathering site in the last week of May. The same is valid for the Southern Florida Acid Rain Test as described in Chapter 4.3.7. In coatings development, such restrictions can delay the start of series production of newly developed coatings by one year or more if outdoor weathering does not begin on time. On the other, the haste associated with the stress of keeping the specified start of outdoor weathering can lead to mistakes in pre-treatment or preparation of the test panels. As sometimes has been observed, test panels did not have the same good quality as areas coated during series production of the end product. The fact that the creator of a new coating can keep his creation under **constant observation during weathering** in his own laboratory is a further advantage of artificial weathering. The ageing behaviour can be monitored without interruption visually or by means of corresponding measuring techniques.

After possible failure of the coating being tested has been recognised, weathering can be immediately interrupted and intervention in the development process of developing can be done. Paint maker and coating user alike have the freedom to **use weathering methods tailored to their products**, in accordance with their own experience and without being restricted or limited by any weathering standards. All weathering stresses and weathering scenarios occurring during outdoor weathering or in normal use respectively can be simulated in the laboratory in principle if effective modern weathering devices are available. The test engineer is free in his decision to accelerate weathering by using stress parameters which may or may not exceed the real stresses. Failure of a weathering device during artificial weathering is not as dramatic as the loss of test panels during outdoor weathering. In other words, **natural catastrophes do not have any influence**.

Modern weathering devices allow weathering stresses to be applied to the test panel surfaces with relatively good local homogeneity. This allows the use of test panels which are much smaller than those used for outdoor weathering. The possibility of **working with small test panels** is also an advantage of artificial weathering. Recall from Chapter 4.6 that salts and dust particles, bird excrement, parts of plants and the like can be heterogeneously deposited on the test panels during outdoor weathering and can lead to special kinds of surfaces damage that hamper assessment of the weathering results. Relatively large test panels are necessary for outdoor weathering in order that such influences may be excluded for the purpose of, for instance, visual assessment or measurement of gloss and colour. Such unwanted **influences can be avoided or studied separately** in artificial weathering. Investigation of each weathering stress factor in isolation from the others is also a key advantage of artificial weathering.

Repeatability and reproducibility, which are still rather poor in artificial weathering, can be expected to improve in future due to continuous improvements in compliance with and monitoring of the specified test conditions. In contrast to outdoor weathering, the repeatability and reproducibility of artificial weathering are not limited by randomly occurring nature-related variations in weathering stress, which are the major cause of poor repeatability of tests performed outdoors.

Let us now take a brief look at the disadvantages of artificial weathering. An important weakness is the fact that the **natural solar radiation cannot be exactly simulated by artificial radiation sources at acceptable cost**. The radiation emitted by artificial sources currently available deviates considerably from the solar radiation. Furthermore, the micro climate on the surface of the coatings to be tested deviates from that in normal use. Weather scenarios randomly occurring in normal use influence the coating ageing to different extents as a function of the coating type, formulation and application. This means that the corresponding worst case weather scenario or scenarios must be met and correctly considered when the test cycles are being designed.

Recommended or specified artificial weathering tests are always only compromises that have limited correlation with practice. Today's highly effective weathering devices are capable of simulating the natural weathering stress to a relatively good approximation in laboratory. However, high performance demands a high price too. **Artificial weathering is rather expensive.** For these reasons, conventional outdoor weathering, which is the cheaper method of accelerated testing, is still practiced in spite of the weaknesses as described in Chapter 4.7.

6.6 Validation in artificial weathering

Simulation of natural weathering stress by means of artificial weathering can only be deemed successful if coatings of different ageing behaviour reveal the same weatherability ranking in the laboratory as they do in normal use. In other words, that coating which proves to be the best in normal use must prove to be the best in artificial weathering and vice versa. Like all other laboratory test methods, artificial weathering must also be validated. According to ISO 17025, validation is the confirmation by examination and the provision of objective evidence that the particular requirements for intended normal use are fulfilled. In the case of artificial weathering, resistance to all natural weathering stresses that occur in intended normal use belong to the aforementioned particular requirements.

Validation is one of the basic prerequisites for the standardisation of test methods. All test methods modified or newly developed by a testing laboratory must be validated by the laboratory itself. Standard validation practice includes the regular use of reference materials, intra-laboratory comparison with results obtained with other weathering methods, inter-laboratory comparisons, systematic assessment of all factors influencing the weathering results, and assessment of uncertainty based on a scientific understanding of the theoretical principles of weathering, and practical experience of the laboratory staff involved in coatings testing. General coatings testing shows that laboratories performing only standardised weathering methods should also take part regularly in inter-laboratory comparison or proficiency-testing programmes.

Intra- and inter-laboratory comparisons are commonly performed on the basis of reference methods. For validation in artificial weathering, it is common to use outdoor weathering as a reference method. In other words, **artificial weathering is validated by outdoor weathering**. This presupposes that the outdoor weathering used for comparison fulfils the particular requirements of a reference method. As discussed in Chapter 4.7, considerable doubts exist about the repeatability and reproducibility of outdoor weathering in general. This has led to the question as to "Which of the outdoor weathering tests performed at which weathering site in which years and with which exposure period should be used as reference?" The answer is that only certain weathering sites, such as the reference climates of Miami, Florida and Phoenix, Arizona, as described in Chapter 4.2.1, can presently be recommended for validation of artificial weathering. If the product specification for the coatings to be tested requires outdoor weathering for 2 years in Miami, Florida, as well as Phoenix, Arizona, both sites must be involved in the validation for the specified exposure period. If one of the two natural reference climates is known to have a weak influence on coating ageing, comparison can be limited to the one exerting the stronger influence.

Three outdoor weathering tests, at least, should be performed consecutively at the same weathering site (reference climate). Whether the average results of the three tests or only the results of the test revealing the strongest damage should be involved is a question of test philosophy. It is common for validation to be done by comparing the weatherability rankings from the artificial weathering method with the outdoor weathering used as reference method. For this purpose, coatings differing in ageing behaviour must be available for the experimental studies. The weatherability of the coat-

$$\text{SPEARMAN} = 1 - \frac{6\sum_{n=1}^{n} D_n^2}{n(n^2-1)}$$

$$\text{PEARSON} = \frac{n(\sum xy) - (\sum x)(\sum y)}{\sqrt{[n\sum x^2 - (\sum x)^2][n\sum y^2 - (\sum y)^2]}}$$

Figure 6.55: Formulas for calculating the ranking correlation coefficients of SPEARMAN and PEARSON (from [95]). n: Number of coatings involved in the weathering test series being compared. D: Difference between the coatings' place numbers inside the test series from the two weathering tests being compared. x und y : Change in the chosen coating property of identical coatings for the one (x) or other (y) weathering test being compared

ings involved in the validation should range from extremely poor to extremely good and form as harmonious a series as possible. Test series should comprise 10 coatings at least. How well the coating rankings from the two methods agree with each other is commonly expressed by the ranking correlation coefficients already shown for some examples in this book. The weathering results dictate whether the methods of assessment by SPEARMAN or PEARSON are used. The formulas for calculating the respective ranking correlation coefficients are shown in Figure 6.55.

If only the ranking place numbers of the coatings are available for calculation, as is common with weathering results assessed by visual methods, the SPEARMAN ranking of correlation coefficients is used. The SPEARMAN method suffers from the disadvantage that adjacent coatings in the ranking which differ by a full place number, irrespective of the extent of the difference in coating damage, are used for the calculation.

If concrete values of the coating property chosen as ageing criterion can be measured, for instance, loss of gloss or change in colour, the PEARSON ranking correlation coefficients (PEARSON in short) tend to be used. PEARSON calculations are also an option of the Microsoft Excel standard software. The formula for calculating PEARSON includes the absolute values, x_i and y_i, of the weathering-induced changes in coating property measured on the two counterparts of the respective coating in the calculation.

The following example taken from weathering practice illustrates differences in PEARSON coefficients. The purpose of this example is to validate four different methods of artificial weathering (methods 1 to 4) by means of the same outdoor weathering, e.g. in Miami, Florida. Ageing criterion is the change in colour. Identical series of automotive coatings are used for each weathering method. Each series contains specimens of the same ten coatings (#01 to #10) displaying sufficient differences in ageing behaviour.

Figure 6.56 presents the validation results in different diagrams for each of the four methods of artificial weathering. In each diagram, the loss of gloss achieved in artificial weathering is plotted against the loss of gloss achieved in outdoor weathering, along with the corresponding PEARSON. To simplify the presentation, the 10 coatings are numbered from #01 to #10 to reflect their weatherability as assessed from outdoor weathering. The artificial weathering methods were chosen for the different extents to which they correlate with practice. Since loss of gloss serves as ageing criterion, objective readings are available and PEARSON can be used to describe the ranking correlation.

Method 1 at the top left corner shows identical loss of gloss for the respective counterparts inside the two respective coating series. In this case, PEARSON = 1. In other words, absolute ranking correlation occurs very seldom in weathering. Cases of the kind shown in top right corner can be found more often. A PEARSON of 0.9 is commonly considered a relatively good correlation in weathering. The identical counterparts inside the respective two coating series differ markedly in their ranking places, though. The trend curve shown in the diagrams is also beginning to level off. In many cases, the test engineer has to live with the situation shown for method 3 in the bottom left diagram. A PEARSON of 0.8 is considered to constitute adequate correlation in weathering.

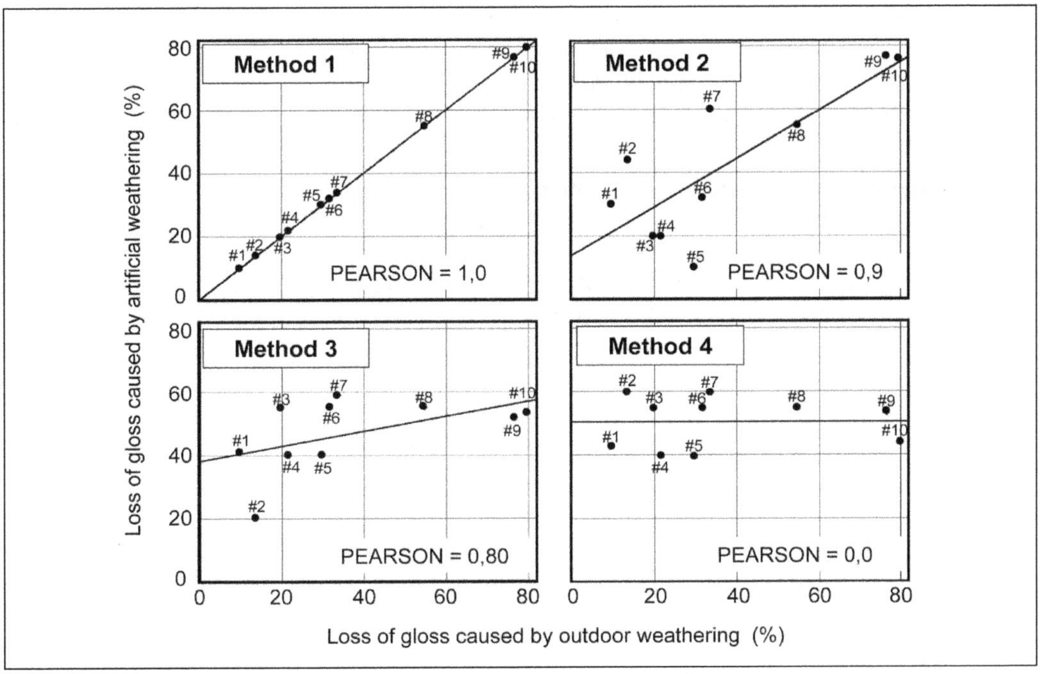

Figure 6.56: PEARSON ranking correlation coefficients demonstrated for a fictitious example wherein four artificial weathering methods (methods 1 to 4) are validated by comparison with outdoor weathering on identical series of ten coatings (#1 to #10) in which of loss of gloss serves as ageing criterion

For values of PEARSON lower than 0.8, as in the case of method 4, the artificial weathering method to be validated fails to meet its goal of simulating natural weathering stress in the laboratory. Consider another hypothetical example to demonstrate the difference between PEARSON and SPEARMAN in assessing ranking correlation.

For this purpose, in addition to the aforementioned example, a fifth method of artificial weathering is compared with the outdoor weathering, which also serves as reference method here. The result is shown in Table 6.11. In this example, the sequences of coatings in the respective columns are identical. However, the loss of gloss increases from coating #1 to coating #10 much less harmoniously in the laboratory than outdoors. In contrast to outdoor weathering, coatings #1 to #8 as well as #9 and #10 differ only very slightly from each other in artificial weathering. A SPEARMAN calculation produces a value of 1 because none of the respective counterparts of the coatings in the two series revealed a different ranking place. A PEARSON calculation yields a value of 0.88, which is less than that of SPEARMAN. This shows that PEARSON is better at reflecting the real conditions.

Table 6.11: Different assessment of ranking correlation by means of different coefficients (SPEARMAN and PEARSON) in the comparison of artificial and outdoor weathering of 10 different coatings

Coating	Loss of gloss (%)	
	outdoor weathering	artificial weathering
#1	10.0	10.0
#2	14.0	11.0
#3	20.0	11.2
#4	22.0	11.3
#5	30.0	11.4
#6	32.0	12.0
#7	34.0	13.0
#8	55.0	14.0
#9	77.0	79.2
#10	80.0	80.0
SPEARMAN	1.00	
PEARSON	0.88	

Ranking correlation can be performed in the same way to reveal differences in weathering stresses between different weathering sites or different weathering periods that arise in outdoor weathering, and differences in methods, exposure periods or weathering devices that arise during artificial weathering. The question as to if, and to what extent, different ageing criteria correlate with each other is often of interest too.

Note that **artificial weathering has only one paragon – nature**. Therefore, methods of artificial weathering that are not validated against corresponding outdoor weathering do not allow correct predictions to be made about the coatings' ageing behaviour in intended normal use. Their results only allow statements to be made only about the coatings' resistance to the weathering stress prevailing in that method of artificial weathering.

7 Ageing criteria used in accelerated testing of coatings

7.1 Phenomenon of photochemical ageing

Figure 7.1 shows the phenomenon of coating ageing that may be caused by natural or artificial weathering for the example of a multilayer automotive coating system, consisting of clearcoat, pigmented basecoat, and filler.

As shown in this Figure, the clearcoat forms the outermost coating layer, and is immediately exposed to all weathering stresses. The appearance of the coating system may change to the same extent, as degradation of the clearcoat progresses. However, degradation of the coating surface mostly does not progress homogeneously, probably due to prehistory-induced heterogeneities inside the clearcoat. Also, certain stresses are observed to heterogeneously attack the coating surface, such as acid dew and biological influences. Irregularly distributed cracks, craters, blisters, and flecks are typical phenomena of photochemical ageing of clearcoats that leads to loss of gloss, increase in haze, and change in colour.

The number, size, and depth of surface defects commonly increase with increasing exposure period, and this makes them more striking. This ageing phenomenon leads to a gradual loss of the initial, mostly highly attractive optical appearance. Increasing surface roughness and possible colour changes promote gradual loss of a coating's intactness and attractiveness. As often observed, weathering leads to the formation of blisters in the interfacial area between clearcoat and basecoat, resulting inevitably in delamination during further exposure. Coatings consisting of more than three layers are especially sensitive to this kind of coating ageing. The reason for this ageing phenomenon is often the insufficient barrier presented by the outermost coating layer, which is mostly only a few micrometers thick, to penetrating UV radiation, oxygen, and water. In such cases, the surface of the subsequent coating layer may not be sufficiently protected against UV. Overly thin layers cannot absorb sufficient stabilisers. Photochemical degradation on interfacial areas often leads to under-chalking, and also spoils the coating's appearance. This

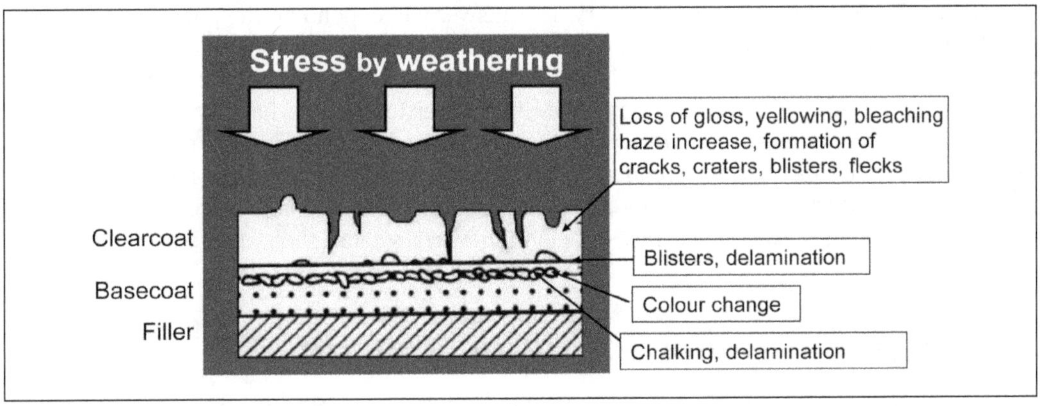

Figure 7.1: Simplified representation of weathering-induced photochemical ageing for the example of a basecoat/clearcoat system

Ulrich Schulz: Accelerated Testing:
Nature and Artificial Weathering in the Coatings Industry

ISBN: 978-3-86630-908-1

effect can become particularly dramatic in coating systems that use black basecoats, leading to striking white spots, as sometimes observed on particularly water-stressed sites of car bodies. UV radiation reaching the basecoat surface is mostly still capable of photochemical ageing. The formation of differently coloured ageing products of the resin or degradation of pigments can lead to a change in colour. Although it is unlikely that UV will penetrate the pigmented basecoat, ageing phenomena caused by additional chemical stress, for instance by acid dew or biological matter, have been observed right down to the substrate. Such effects are more often observed with one-coat paints. In these cases, weathering can lead to corrosion phenomena on the substrate, even if the coating is not intentionally damaged by deep scratches, as is common in corrosion tests described in Chapter 2.2.

Modern high-gloss automotive coatings are also subjected to certain mechanical stresses during normal use. Such stresses mostly come from the natural environment, such as impact by stones, gravel, and airborne sand, contact with building materials, branches, keys, and fingernails, or car-washes. Recent publications have shown that coatings with good mechanical resistance in their freshly cured state gradually lose their initial excellent mechanical properties with increasing duration of natural or artificial weathering. Therefore, the influence of weathering on the coatings' resistance to mechanical stresses must be considered as well.

The coating user should not light-heartedly eschew monitoring the action of weathering on important mechanical properties. Customers expect that the coatings of the cars they have bought will not lose their good new car appearance due to loss of mechanical resistance after only a short time of normal use. As already mentioned in Chapter 2.3, regular use of commercially available car-washes is now considered normal use of automotive coatings. It is also known that the initially good elasticity, as well as the capability of clearcoats to heal mechanical defects themselves, is progressively lost with increasing stress by weathering, too. Changes in adhesive strength, stone impact resistance, and scratch resistance should be monitored unless it has been found that they change harmoniously to each other during weathering. With pigmented one-coat paints, chalking, as already described in more detail in Chapter 3.1.4, must additionally be considered an important ageing phenomenon.

7.2 General thoughts on choosing ageing criteria

The test engineer involved in coating weathering will ask whether all the aforementioned ageing criteria must be monitored in his laboratory. The answer will depend on the purpose of his task. Monitoring current series production requires that only a certain selection of ageing criteria be monitored. This selection must allow adequate assessing of compliance of the coatings' quality concerning their ageing behaviour. Some coating properties change harmoniously with each other. Others show no negative alterations in normal use at all, and need not be monitored during weathering. Monitoring of series production can therefore mostly be limited to just a few coating properties. In contrast, as many coating properties as possible should be involved in the development of new coating creations because sufficient knowledge about their ageing behaviour is for most part unavailable.

7.3 Aspects of selecting ageing criteria

Almost all the coating properties that might be of interest in assessing the coatings' ageing behaviour could be monitored during weathering with the modern commercially available technical means. The price of appropriate measuring instruments ranges from several hundred to several millions of Euros. The desire to monitor weathering-induced changes as thoroughly as possible mostly fails due to lack of financial resources. Therefore, prior to purchasing measuring equipment, the test laboratory will always have to confront the question: "How many ageing criteria need to be chosen to assess the coatings' ageing behaviour adequately?" The answer must consider the aspects of reasonable scope, economic acceptability, sufficient accuracy, and reliable service life prediction.

Note that the chosen ageing criteria must provide enough arguments to convince everyone involved in developing and using the coatings to be tested of the outcome of the assessment (i.e., good or bad weatherability). Economic aspects now need some consideration. An assessment, for instance, of the intactness of a coated area is predominantly influenced by its optical state as perceived by the human eye. Therefore, evaluation of the results of accelerated testing must be oriented towards the visual impression as an important ageing criterion. As an example, the coating user will not be particularly impressed by information that the carbonyl index of his coating will increase markedly during the guaranty period if the appearance still fulfils his expectations. However, he will complain if a high-gloss coating develops matt spots, colour anomalies or too many scratches after a relatively short time of normal use, despite negligible loss of light stabilisers.

As this example shows, the simplest ageing criterion is the intactness of the coating surface and uses the cheapest method for assessing weathering results, namely evaluation with the naked eye. However, statements on the basis of this method are severely limited. In the simplest case, the only judgement possible is "the surface is or is not intact". Better graduation can be achieved, for instance, by comparison with appropriately graduated test panels that have been previously weathered. Instead of real test panels, reference paper pictures can be used for assessing the degree of weathering-induced damage. Graduation by means of appropriate reference pictures is often used for inter-laboratory comparison, too. A typical example is the determination of intensity, quantity and size of common types of defects according to ISO 4628 [92]. The standard contains appropriate reference pictures suitable for almost all types of weathering-induced surface defects.

A cheap method also exists for evaluating the ageing criterion "change in colour". It uses the grey scale standardised in ISO 105-A2 [93]. The grey scale is a simple system for visually evaluating the colour difference between two test panels by comparison with differences between two greys. Figure 7.2 shows its simplicity on the one hand, but on the other hand also reveals the weakness of this typically subjective method, with its high uncertainty. In particular, transferring the colour difference between the two weathered test panels to one of the standardised levels of the grey scale is heavily influenced by the person performing the evaluation.

Testing laboratories intending to employ good laboratory practice or achieve accreditation to ISO 17025 can hardly get by on such simple and cost-saving methods of evaluation.

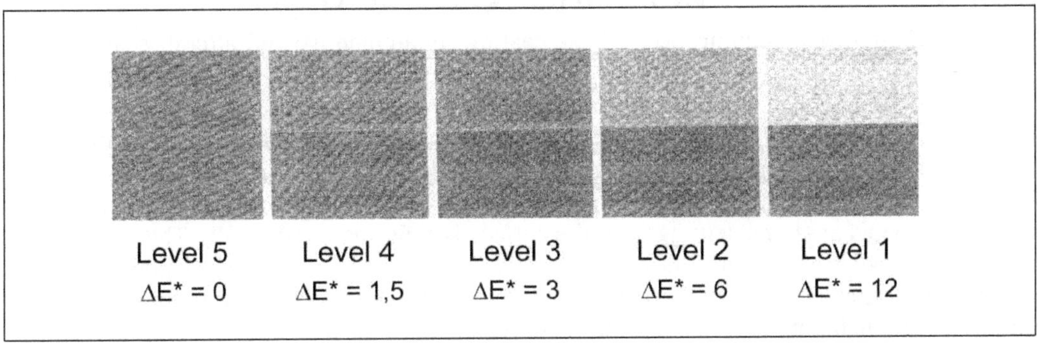

*Figure 7.2: Standardised levels on the grey scale, according to ISO 105-A2, with information about the corresponding ΔE*values (from [42])*

Table 7.1: Selection of methods for measuring important coating properties suitable for monitoring state of ageing during weathering

Ageing criterion	Standard (examples)	Assess-ment
loss of gloss (20°, 60°, 85°)	DIN EN ISO 2813	objective
haze increase	DIN EN ISO 13803	objective
change of colour	ISO 7724	objective
blistering	DIN EN ISO 4628 T 1	subjective
rusting	DIN EN ISO 4628 T 2	subjective
cracking	DIN EN ISO 4628 T 3	subjective
flaking (delamination)	DIN EN ISO 4628 T 5	subjective
chalking	DIN EN ISO 4628 T 6 & 7	subjective
adhesive strength (cross-cut test)	DIN EN ISO 2409	subjective
scratch and mar resistance	DIN EN ISO 11998 E DIN 55668	objective

Table 7.1 shows a selection of methods for monitoring ageing criteria during weathering and for assessing the ageing behaviour. They are often used in relation to accelerated testing. As can be seen, a lot of the methods are based on visual comparisons. As mentioned before, the results of such methods are always characterised by considerable uncertainty. Appropriate efforts are necessary to keep the uncertainty down to acceptable limits, for instance, by frequent participation in inter-laboratory comparisons.

The uncertainty of the aforementioned subjective methods can also be limited if accompanied by appropriate optical means, such as microscopes or macroscopes that allow surface photographs to be taken. Although associated with higher costs, macroscopes employing co-axial or other special kinds of illumination are state of the art and can be found in efficient test laboratories. If used to support visual evaluations, the reason for surface defects can be better identified. Only in this way, for instance, is it possible to distinguish weathering-induced defects clearly from surface defects caused by improper handling of the test panels. Surface photographs demonstrating the weathering-induced surface damage afford a good means of qualifying test reports, and allow lasting archiving of the test results. In some cases, such as monitoring of series production, subjective evaluation methods may suffice. However, note that for testing the coating's quality during the process of development or in the selection of coatings offered by different paint makers, subjective methods should be complemented by appropriate objective ones. The subjective error that arises, for instance, when the opinions of inspectors evaluating the coating damage drift from each other may be too large to ensure reliable estimation.

7.4 Ageing criteria for high-gloss coatings

In view of their special significance, vehicle coatings of automotive and aircraft coatings will predominantly serve as examples below. With a few exceptions, such as camouflage coatings for military cars, vehicle coatings are mostly characterised by high gloss. Today's automotive coatings are predominantly applied in the form of multilayer systems. In this case, the clearcoat, applied as the topmost layer, is most responsible for the high optical quality and appearance of the car body. A key property of high gloss is its specularity. In other words, mirror images are clearly recognizable in the coated area. Incident light on the surface is reflected directionally. This means that the angle of incidence is equal to the angle of reflection. Weathering-induced formation of cracks, craters, blisters, mat flecks, and other surface defects, leads to increasing roughness. Rough surfaces reflect light, not only in the principal direction, but also diffusely in all other directions. The appearance's quality decreases. An object reflected by the coating surface no longer appears brilliant, but only blurred.

Second-hand cars mostly have a matte finish. This has led to the question: "Which ageing criterion can best reflect this kind of surface damage in high-gloss coatings?" Loss of gloss is a good measure for assessing the weathering-induced decrease in a coating's appearance and, therefore, is predominantly used as an ageing criterion in coating weathering. For this purpose, the directionally reflected light is measured. A wide range of measuring instruments is available in all kinds of efficiency and price ranges. For best effect, different angles of incidence for measurement are recommended. High-gloss coatings (e.g. automotive coatings), should be measured at 20°, medium-gloss coatings (aircraft coatings) at 60°, and low-gloss coatings at 85°. Common glossmeters, often called reflectometers, must be regularly calibrated by means of gloss standards, mostly made from black-coloured, highly polished glass. A gloss standard adjusted to about 100 gloss units (GU) at 20° corresponds quite well with the reflectance of common high-gloss automotive coatings (mostly showing about 90 GU). Therefore, further discussion of gloss will concern data measured at 20° unless otherwise stated. In some cases, it makes sense to track reflection haze for high-gloss coatings, too. High values of haze manifest themselves in a blurry mirror image. Modern glossmeters offer haze measurement as a complementary option. For measuring the haze, the intensity of diffusely reflected light is measured over an angle range of 1.8° around the 20° base. Figure 7.3 shows the principle of gloss and haze measurement.

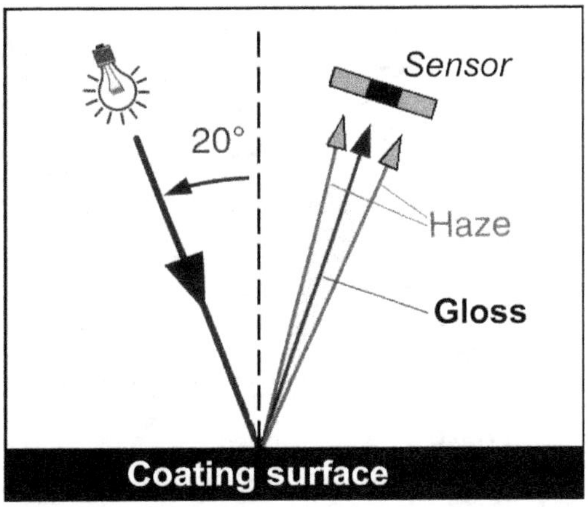

Figure 7.3: Principle behind the measurement of gloss 20 and haze, showing the portions of the reflected light measured as gloss (black arrow on the right) and measured as haze (grey arrows on the right), adapted from [92]

7.5 Haze or gloss as ageing criterion

Haze often responds more sensitively than gloss to certain forms of surface damage, for instance, acid etching during weathering in acid-stressed environment. Although they influence the judgement during visual assessment, acid etching often cannot be evidenced by changes in gloss. Also, coatings damaged by the typical pinholes caused by biological attack can be identified better by haze increase than by loss of gloss. In all cases, where the loss of gloss does not allow reliable differentiation between weathered coatings, attempts should be made to use haze increase as a substitute. Figure 7.4 shows good ranking correlation between Haze increase and visually assessed degree of damage of a series of 20 modern basecoat/clearcoat systems weathered outdoors in Jacksonville, Florida.

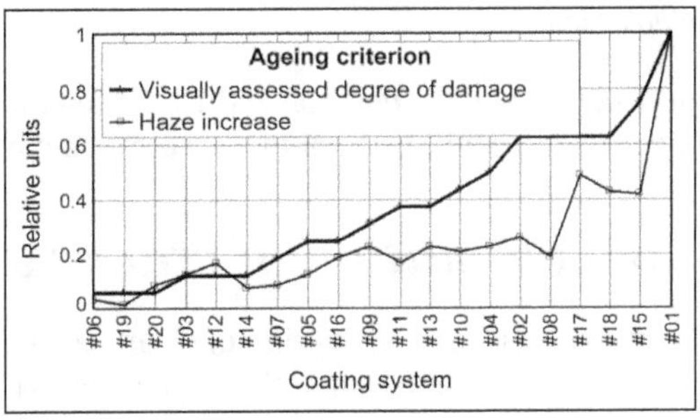

Figure 7.4: Visually assessed degree of damage in comparison to the objectively measured haze increase of a series of 20 different automotive coatings (#01 to #20) weathered outdoors in Jacksonville, Florida, over 14 summer weeks, adapted from [84]

In order to gain reproducible readings, absolute compliance concerning measurement geometry is very important in reflectometry. Gloss and haze cannot be correctly measured on test panels that have been distorted even by only a very small amount during weathering or handling. As reported in [96], changes in the surface profile caused by inhomogeneous shrinking or swelling during wetting and draying may influence gloss measurement. Therefore, care must be taken during preliminary

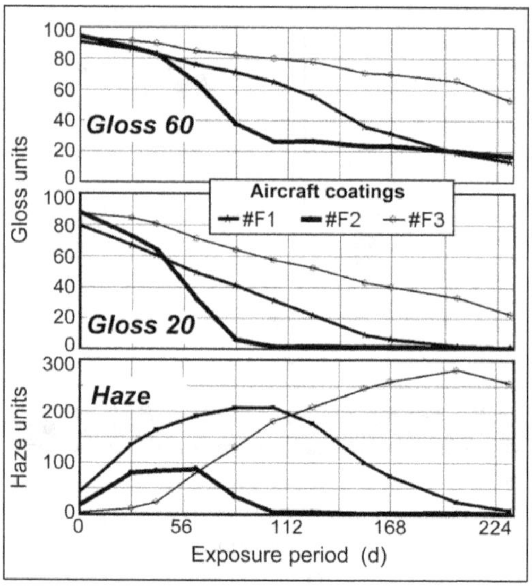

Figure 7.5: Gloss 60, gloss 20 und haze in the course of the ADF test (type M2.5) carried out in a UV2000 fluorescent device for a series of three white coloured aircraft coatings (#F1, #F2 and #F3)

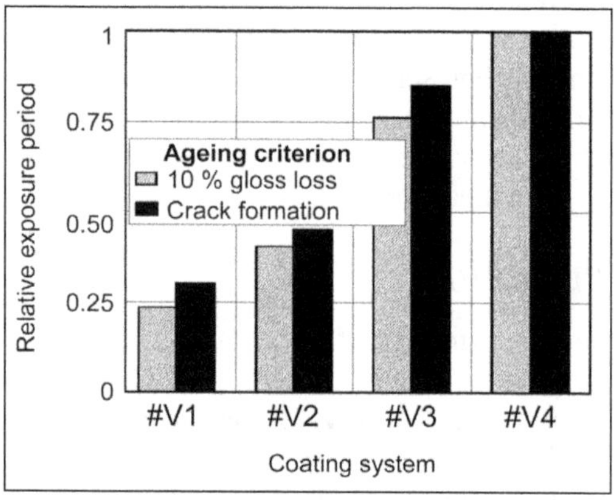

Figure 7.6: Relationship between weathering-induced cracking and loss of gloss for a series of four differently stabilised samples of a UV curable clearcoat, adapted from [16]

cleaning procedures that employ chemicals such as water or polish. In outdoor weathering, consideration should be accorded to whether the test panels are taken from the weathering rack after a longer rainy or after a longer dry phase. Gloss and haze should always be measured after identical pre-treatment.

In the context of reflectometry, the question remains as to "How can haze increase and loss of gloss be harmonised with each other?" Figure 7.5 answers this question for the example of a series of three typical aircraft coatings. Gloss 20, gloss 60 and haze were monitored in an ADF test (type M2.5 as described in Chapter 6.2.4), and are presented as a function of exposure period in separate charts. As can be seen, gloss and haze behave differently in the course of weathering. While the gloss curves gradually approach the value 0, the haze curves pass through a maximum. Furthermore, all the three optical properties correlate well with each other.

First, let us compare the gloss curves for the different measuring geometries. No substantial differences are evident between the curves for gloss 60 and gloss 20. Nevertheless, gloss 20 proves to be the better measuring geometry in this case, because the greatest selectivity was reached at shorter exposure.

Now, let us compare gloss and haze. Clearly, haze attains its best selectivity after 28 days of exposure, faster than both types of gloss. This shows that haze responded faster to weathering stress. The use of haze increase as an ageing criterion allows the exposure period to be shortened without influencing the correlation. Good correlation between the positions of the gloss curves and the position of the haze maximum is also apparent. However, the haze maximum is mostly not achieved in outdoor weathering or in artificial weathering without acid stress.

7.6 Formation of cracks as ageing criterion

As shown in Figure 7.1, ageing of clearcoats can manifest itself in the formation of cracks of different depths. Evaluation of this ageing criterion is almost exclusively of the visual kind, mostly by use of optical means that allows appropriate magnification, as already shown in Figure 6.28 for the example of the "BASF clearcoat crack standard". Crack formation is no longer considered a problem with today's high-gloss automotive coatings. Nevertheless, if cracks occur, the coatings concerned differ in the time to crack formation, mostly in the same order as loss of gloss. Figure 7.6 shows a typical example. The exposure period up to the formation of cracks, and the exposure period up to 10 % loss of gloss are shown together for four coating systems whose clearcoats were stabilised in different ways. This presentation shows an absolute correlation between the two ageing criteria. In this case, preference should be given to that ageing criterion which allows better and faster selection of the coatings to be tested.

7.7 Change in colour as ageing criterion

Let us now look at ageing criterion of change in colour. As mentioned before, this is one of the most widespread ageing criteria in coatings testing, after loss of gloss. For instance, the purchaser of an automobile expects the colour of his car, which is mostly selected to suit his special wishes, to have high longevity. Furthermore, he expects the ageing behaviour to be identical with that of other car body parts which might need painting or replacement following minor accidents. This means that the parts painted afterwards or replaced must not change their colour at a rate different from the original parts during the further normal use. These aspects underline the importance of colour as an ageing criterion.

As already mentioned, different coating components as well as different coating layers can be responsible for loss of gloss and change in colour. Therefore, loss of gloss and change in colour will mostly behave differently in the course of weathering. As shown in Figure 6.12, the two ageing criteria mostly do not progress harmoniously. This is often observed in the case of multi-layer coating systems.

Subjective as well as objective methods for measuring change in colour are common. Visual evaluation of weathered test panels against the corresponding non-weathered ones often leads only to the assessment "change/no change", possibly with the supplement "more/less pronounced". Also, the aforementioned grey scale for visual assessment of colour differences is often unsuitable for attaining the selectivity necessary for sufficiently differentiating the weathered test panels. Therefore, subjective colour comparisons have mostly been replaced by objective measuring instruments. Nowadays, effective testing laboratories are equipped with appropriate colorimeters, which allow objective measurements. The human eye can differentiate between about 10 million hues. Therefore, colorimeters must have very high selectivity, and this leads to relatively high prices. The use of standardised methods has been approved. The CIELab scheme described by [94] is recommended by the CIE and is the mostly employed scheme in colorimetrics.

Simply explained and shown in Figure 7.7, the colour scheme consists of the two chromaticity coordinates, a* (green-red) and b* (blue-yellow), which define the hue of the colour. These two coordinates are arranged at right-angles to each other on a fictitious horizontal colour plane. A third coordinate, L*, defines the lightness value. L* is arranged at right-angles to the plane formed by a* and b* and characterises greying as well as lightening of the coating surface. Colorimeters are capable of displaying the changes in the values of the individual coordinates Δa^*, Δb^* or ΔL^*, as well as the whole change in colour, ΔE^*, calculated as shown in Figure 7.7. ΔE^* is mostly used to express weathering-induced change in colour. In coatings that tend only to yellowing or bleaching, for instance, basecoat/clearcoat systems with white basecoats, the value of Δb^* is often used to characterise the change in colour. It has become common to use Δb^* instead of the change in the yellowness index, ΔYI, as defined in [12].

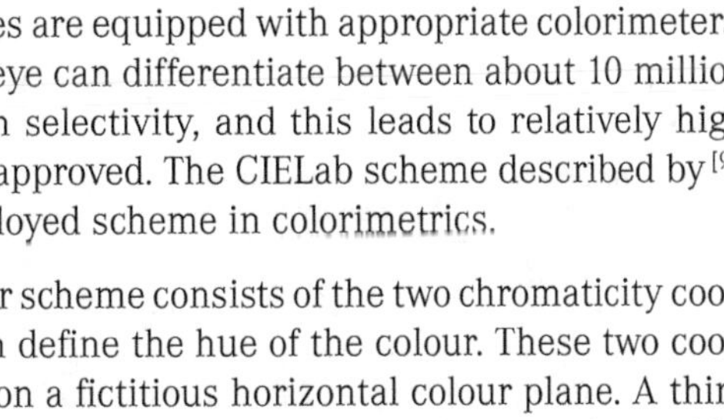

$$\Delta E^* = \sqrt{(\Delta L^*)^2 + (\Delta a^*)^2 + (\Delta b^*)^2}$$

Figure 7.7: CIELab colour scheme, from [92]

7.8 Adhesive strength as ageing criterion

As discussed in Chapter 7.1, adhesive strength can also be a key ageing criterion. Weathering-induced changes in adhesive strength cannot be excluded. Many methods have been developed for this purpose, as described in [97]. The cross-cut test is the most prevalent in coatings testing.

Figure 7.8 shows the surface of a grey-coloured, one-coat paint on a white-coloured plastic substrate after the cross-cut test. The tested panel is compared with the corresponding reference photograph standardised in DIN ISO 2409 [98], and shown here in the top right corner. In this example, for instance, comparison yields the standardised value of 3 for characterising the adhesive strength. In spite of its simplicity and highly subjective character in the performance and evaluation, the results of the cross-cut test are astonishingly highly repeatable and reproducible. This is borne out by inter-laboratory comparisons. Since the effect of accelerated testing is best expressed by relative values, including the cross-cut test, absolute values are not the primary focus. The use of relative values for change in adhesive strength can keep the uncertainty down. If it is necessary to track the weathering-induced changes in adhesive strength, the cross-cut test can be recommended as a simple and representative method.

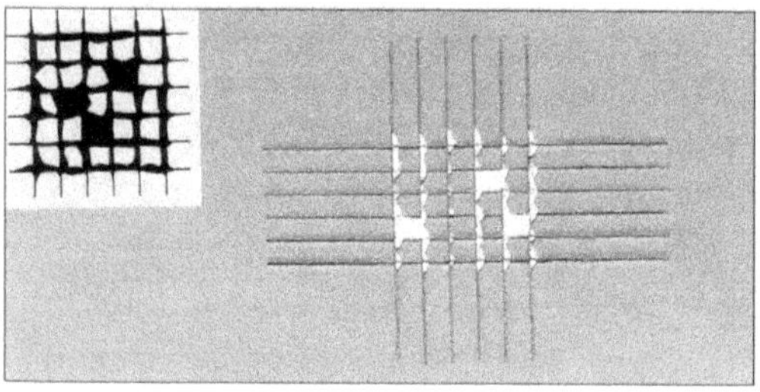

Figure 7.8: Photograph of the surface of a grey-coloured top coat on a white plastic substrate after the cross-cut test in comparison with the respective reference photograph (top left corner), according to DIN ISO 2409, adapted from [94]

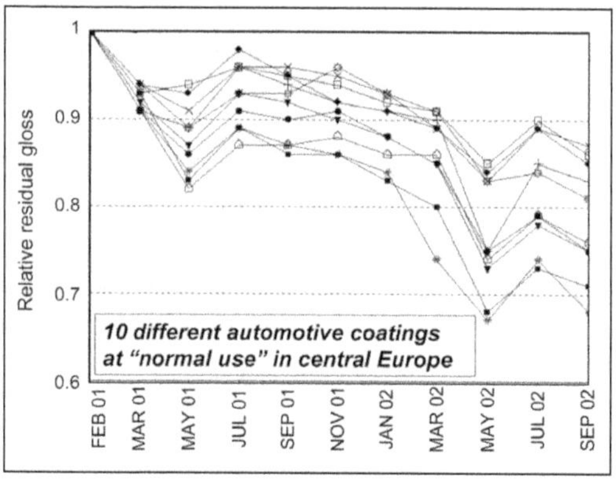

Figure 7.9: Gloss in the course of normal use including the use of real car-washes (once a week from February 2001 to September 2002) for a series of ten different automotive coatings, adapted from [99]

Figure 7.10: Photograph of the Amtec-Kistler laboratory car-wash for determining car-wash resistance

7.9 Scratch resistance as ageing criterion

As mentioned before, scratch resistance is an important coating property. The clearcoat, which is the most responsible for this property, undergoes changes in mechanical strength by crosslinking and chain scission during weathering. This can lead to considerable changes in scratch resistance. Weathering-induced decrease in scratch resistance can considerably accelerate spoiling of the coating's initial, usually excellent appearance. The studies discussed below clearly revealed the influence of weathering-induced changes in scratch resistance on the coating's appearance during real normal use. In other words, the coatings were exposed to weathering stress combined with typical mechanical stress. As described in [99], a test series consisting of ten automotive coatings was stressed in the same manner as a car roof in normal use. To this end, the coatings were stressed mechanically by regular use of real car-washes, as well as thermally and photochemically by the natural environment. Clearcoats were chosen for their almost identical weatherability but different scratch resistance for this experiment. This real practice test ran for 19 months in central Europe. For this purpose, appropriate test panels were fixed on the roof of an automobile that was driven, parked and washed as normal. Washing was carried out once a week in common car-washes equipped with brushes. The gloss of the test panels was monitored throughout the exposure period and is presented in Figure 7.9.

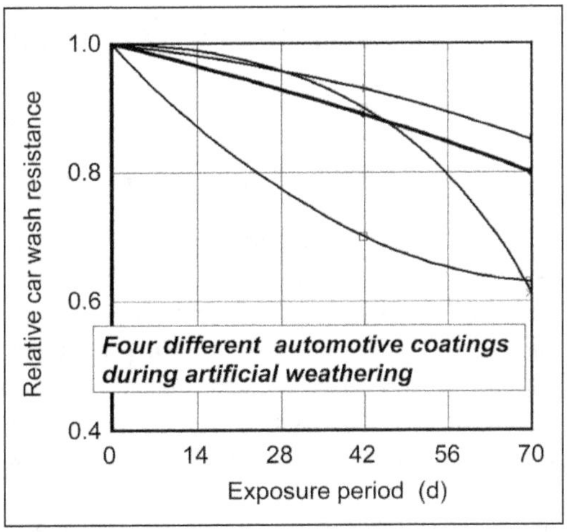

Figure 7.11: Car-wash resistance of four typical automotive coatings in the course of artificial weathering, in a fluorescent UV lamp device, determined by means of the AMTEC KISTLER laboratory car-wash, adapted from [99]

The gloss curves reveal the complex action of mechanical, thermal, chemical, and photochemical stresses that act in normal use. A gradual decrease in gloss was expected due to the gradually increasing number of car-wash scratches. However, the curves show interim occurring increases in gloss from May to July in 2001 and 2002. Since the test was carried out in central Europe, the effect of reflow obviously seemed to be more predominant than the photodegradation effect in the high-temperature seasons, in contrast to the other month. Recall that reflow is the capability of the clearcoat to heal scratches itself if stressed by elevated temperature. Obviously, the test panel surface reached a temperature high enough to trigger reflow during irradiation in the summer months. However, the curves also show that this positive coating property was lost as a function of coating type, due to photochemical ageing, with increase in duration of exposure. The increasing number of remaining car-wash scratches seems to be the reason for the steady decrease in gloss. When the test concluded, the markedly differing gloss values revealed enormous weathering-induced differentiation of the coatings as regards their scratch resistance.

In accelerated testing of automotive coatings, scratch resistance should be included as an important ageing criterion. A suitable method for determining the scratch behaviour of automotive coatings is described in E DIN 55668 [100]. The laboratory car-wash from AMTEC-KISTLER, shown in Figure 7.10, is recommended in this standard as a suitable measuring technique. This instrument uses a brush which is almost identical in diameter with that used in real car-washes.

Figure 7.11 shows the influence of artificial weathering on car-wash resistance, as determined by means of the Amtec-Kistler technique, for four coatings that exhibited the best scratch resistance in the aforementioned practice test. The loss of scratch resistance of up to 40 %, which can be seen in this Figure after an exposure period of 70 days, is also evidence of the need to include this coating property in ageing criteria employing the Amtec-Kistler or some other technique.

8 Standards for accelerated testing

8.1 Organisations for standardisation

Standardisation is an instrument for regularising the technical world in general. Technical standardisation is performed at national and international levels. DIN ("Deutsches Institut für Normung") is responsible for national legislation in Germany, CEN ("Comité Européen de Normalisation") is responsible for Europe standards, while ISO ("International Organisation for Standardisation") looks after international standards. Test standards, in general, and weathering standards, in particular, have a particular meaning. The purpose of weathering standards is to regularise accelerated testing with a view to achieving conformity in this field of testing at national and international level. Conformity means, in this case, that outdoor weathering sites with identical climate or identically equipped testing laboratories are capable of providing comparable weathering results. In other words, identical coatings should experience the same degree of ageing if tested on the basis of the same weathering standard, irrespective of where the test is performed, for instance, whether in the USA or Germany.

Figure 8.1 shows the hierarchy, relationships and demarcation lines of the most important international standards for outdoor weathering, accelerated outdoor weathering by means of radiation concentration, and artificial weathering. The weathering standards in the three boxes for polymers in general form the basis for the standards at the next lower level for coatings in particular. Thus, DIN EN ISO 2810 (outdoor weathering), DIN EN ISO 11341 (artificial weathering employing xenon lamps), and DIN EN ISO 11507 (artificial weathering employing fluorescent UV lamps) are the weathering standards specially tailored for coatings weathering.

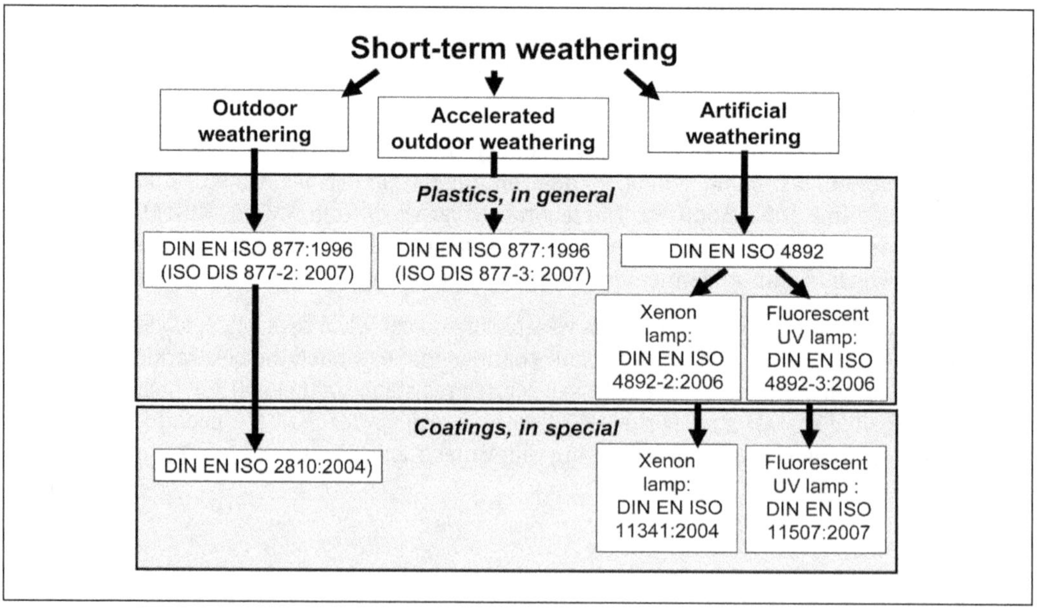

Figure 8.1: Hierarchy of international standards relating to short term weathering of polymers, in general, and organic coatings, in particular

Ulrich Schulz: Accelerated Testing:
Nature and Artificial Weathering in the Coatings Industry

ISBN: 978-3-86630-908-1

8.2 *Test specifications of companies and industrial associations*

Many test specifications for coatings have been developed separately for special purposes, paint makers, and coating users. These specifications are almost exclusively published by companies or associations of companies that produce or use coatings, for instance, the "Verband der Automobilindustrie" VDA in Germany or the Society of Automotive Engineers SAE in America.

They commonly contain regulations for testing the resistance to many stresses from the environment. Requirements in terms of accelerated testing have fortunately for the most part been taken over from the over-riding international standards listed in Figure 8.1, and mostly modified to a minor extent for their special purpose.

Weathering standards are recommendations, like all other standards. However, they are obligatory for testing laboratories which offer testing service in accordance with concrete test standards. Testing in absolute agreement with a standardised or otherwise specified test method has the important advantage for the laboratory that it is not responsible for deviations between the test results and the coating's performance in real normal use. In this case, the testing laboratory cannot be held responsible for complaints, for instance, by car users seeking recoating of their cars. Of course, the testing laboratory performing weathering service should advise the customer if it recognises serious flaws in the desired test specification. If the customer insists on what is a flawed test method, the test report must expressly declare that the test results are only valid for the test conditions employed and are unsuitable for predicting the coating's service life in the intended normal use. In such cases, the testing laboratory cannot be sued for damages arising from poor coating performance.

However, any testing laboratory that claims to be competent is responsible for performing the test to a good quality standard in compliance with the respective test standard. The duties and obligations of a testing laboratory are regulated in ISO 17025 "General requirements for the competence of testing and calibration laboratories". Accreditation on the basis of this standard affords a good means of attaining and maintaining high-quality test results.

8.3 *Further development of test standards*

Test standards are said to positively and negatively influence the state of the art, especially of weathering technology. As already mentioned, standards always lag behind current know-how. Where standards are revised to include the current scientific knowledge, the demand for standards-compliant testing necessarily means that the corresponding weathering technology must be available. This can promote the development and production of improved weathering devices. If, on the other hand, test standards focus only on developing to the current, partly antiquated state of weathering technology, for instance, with preference accorded to economics aspects or certain company interests, there will be no demand for new or advanced weathering devices. In this case, weathering technology will lag behind the well-founded demands for improved prediction of coatings service life, as discussed later in connection with Part 3 of ISO 4892.

In general, test standards are the intellectual product of the persons involved in their creation or development. General experience shows that the creation or development of standards is mostly time consuming. Therefore, **test standards reflect yesterday's know-how**. As is often observed, and this also applies to coating weathering, new experiences in the special field of accelerated testing or the introduction of coatings based on new materials raise doubts concerning the sense or correctness of weathering standards that currently exist. However, not only the time-consuming process of creating of new standards can be seen as a reason for possible discrepancies relative to current know-how in weathering, but also commercial or other interests. Consideration and

Table 8.1: Standardised exposure cycles for artificial weathering using filtered xenon arc radiation according to DIN EN ISO 4892-2, Method A [101]

Method A Artificial weathering using xenon lamps with daylight filters						
cycle		irradiance (continuous)		BST*	CT**	relative humidity
Cycle No.	exposure period	broadband (300 to 400 nm) W/m²	narrowband (340 nm) W/m²/nm	°C	°C	% R.H.
1	102 min dry	60±2	0,51±0,02	65±3	38±3	50±10
	18 min water spray	60±2	0,51±0,02	-	-	-
2	102 min dry	60±2	0,51±0,02	65±3	not controlled	not controlled
	18 min water spray	60±2	0,51±0,02	-	-	-
3	102 min dry	60±2	0,51±0,02	100±3	65±3	20±10
	18 min water spray	60±2	0,51±0,02	-	-	-
4	102 min dry	60±2	0,51±0,02	100±3	not controlled	not controlled
	18 min water spray	60±2	0,51±0,02	-		-

** Black standard temperature ** Chamber temperature*

ignoring of current know-how are readily demonstrated for the purposes of the present discussion by Parts 2 and 3 of the standard ISO 4892, "Plastics - Methods of exposure to laboratory light sources", both published in 2006. Although, intended for plastics in general, ISO 4892 is to serve as the over-riding document for ISO 11341 and ISO 11507, which are the corresponding weathering standards for coatings.

First, let us have a look at the newly developed Part 2 "Xenon-arc lamps. Method A, Artificial weathering using xenon lamps with daylight filters". This part of ISO 4892 is a good example of how know-how was properly included in the development of a standard for artificial weathering. As presented in simplified form in Table 8.1, four exposure cycles are recommended for testing the weatherability.

The first thing to note is that all four exposure cycles employ continuous irradiation. **Avoiding dark phases during artificial weathering reflects the latest experience gained in polymer testing that exposure without irradiation can be seen as a waste of time.** Employing continuous irradiation is a suitable means of accelerating coating ageing and, therefore, enhancing the efficiency of accelerated testing.

Second, the experience that the presence of **UV radiation of unnaturally high photon energy diminished the correlation between the weathering results and practice**, is reflected in the requirement that only a very small part, maximal 0.15 %, of the xenon arc radiation reaching the test panel surface, may be below 290 nm.

Third, there are clear attempts at better simulating the outdoor weathering performed in southern Florida. Two years' outdoor weathering at this site are required nowadays for service life predictions for automotive coatings. As discussed in Chapter 6.2.2.3, **xenon lamp devices allow very good simulation of weather scenarios characterised by direct irradiation**. Recall Chapter 3.2.5, where the coating surface micro climate differs markedly from the climate of the ambient atmosphere during such weather scenarios. Higher temperature and correspondingly lower relative humidity occurs on the coating surface. As also mentioned several times before, weather scenarios involving direct irradiation are only one of the possible worst-case weather scenarios for coatings in normal use. The climatic test parameters specified for exposure cycle 1 (BST = 65 °C, chamber temperature = 38 °C, and relative humidity of the chamber atmosphere = 50 %), correspond quite well with the weather conditions that can be found on horizontally oriented test panels under clear skies in Miami, Florida, at noon in June (see Figure 3.46). This means that exposure cycle 1 meets the natural climatic extremes during irradiation with direct sunlight. When a worst-case weather scenario is applied at a higher frequency than occurs outdoors, in combination with the natural extremes of the weathering stresses, as was done in this case, exposure cycle 1 meets the requirements for a typical time-compressed test method. In this example, development of a standard for artificial weathering was done sensibly and reflected current know-how and all the possibilities of a modern xenon lamp device.

Now let us look at the newly developed Part 3 of ISO 4892, “Artificial weathering using fluorescent UV lamps, Method A”. The exposure cycles are presented in Table 8.2. As known, natural irradiation only by diffuse radiation, which frequently occurs under partly clouded skies, can also be one of coatings’ worst-case weather scenarios. Such a weather scenario can occur when the sun is covered by a cloud under partly clouded skies during outdoor exposure. It is characterised by the coating surface micro climate’s being almost identical with that of the ambient atmosphere, reaching as much as 35 °C and almost 100 % relative humidity, for instance, in Miami, Florida, at noon in June. As discussed in Chapter 3.2.3, water in gaseous and liquid form, mostly boosts coating ageing enormously during irradiation. As described in more detail in Chapter 6.2.2.3, such a weather scenario cannot be reasonably simulated in a xenon lamp device because of the presence of a large quantity of VIS and NIR. Recall that such a natural climate combination of UV radiation and extremely high relative humidity immediately on the coating surface at a subtropical ambient temperature can only be reasonably simulated in a fluorescent UV lamp device. As already shown in Figure 6.11, the climatic working and controlling range, for instance, of the GUT200 covers this climate combination.

Artificial weathering carried out in parallel on two test series or intermittently on one test series (for instance, changing weekly from device to device), with a xenon lamp and a fluorescent UV lamp device, both in proper working order, correctly and completely simulated and accelerated the weathering stress occurring in Miami, Florida. According to the current know-how in the field of weathering, both of the possible worst-case weather scenarios could be involved in artificial weathering in this case.

Let us return to the exposure cycles standardised for artificial weathering using fluorescent UV lamps, as presented in Table 8.2. Although cautiously considering the advantages of continuous irradiation, and wetting by spraying, unfortunately, none of the four exposure cycles reflects current know-how of the significance of water during weathering. All four exposure cycles employ uncontrolled (cycles 1 and 2) or very low (cycles 3 and 4, with <15 %) relative humidity during the dry phases that occupy the most exposure time. Leaving such a very important test parameter uncontrolled is inexplicable. Besides giving away an essential means of accelerating coating ageing and shortening the necessary exposure period, uncontrolled relative humidity can negatively influence repeatability and reproducibility of the weathering results. Furthermore, devoting an own method © and an exposure cycle (6) using the fluorescent UV lamp UV-B-313, which is known to emit a substantial amount of radiation <300 nm, contradicts current weathering know-how. In contrast, a warning should be given about this lamp, at least, in the form of an appropriate note in the standard’s description.

What about the technical possibilities of performing an exposure cycle that corresponds to the appropriate worst case weather scenario from Miami, Florida, under diffuse irradiation? An exposure cycle characterised by 40 °C/95 % R.H. during the dry phase, water spray during the wet phase (1 h every 6 h) and continuous irradiation with an efficient fluorescent UV lamp device, such as the GUT200, fitted only with UV-A-340 lamps, could be recommended to best simulate the natural environmental stresses according to the present state of the art.

Table 8.2: Standardised exposure cycles for artificial weathering using fluorescent UV radiation according to DIN EN ISO 4892-3, Method A [102]

Method A Artificial weathering using fluorescent UV lamps					
cycle		lamp type	irradiance	BST	relative humidity
Nr.	exposure period				
1	8 h dry	UV-A-340	0.76 W/m²/nm at 340 nm	(60 ± 3) °C	not controlled
	4 h condensation		dark	(50 ± 3) °C	not controlled
2	8 h dry 0.25 h water spray 3.75 h condensation	UV-A-340	0.76 W/m²/nm at 340 nm dark	(50 ± 3) °C Not controlled (50±3) °C	not controlled not controlled not controlled
3	5 h dry	lamp combination	continuously 45 W/m²	(50±3) °C	<15 %
	1 h water spray		(290 to 400 nm)	(25±3) °C	not controlled
4	5 h dry	lamp combination	continuously 45 W/m²	(70±3) °C	<15%
	1 h water spray		(290 to 400 nm)	(25±3) °C	not controlled

From the discussion of the newly developed Part 3 of ISO 4892, the following question inevitably arises: "What is the reason for almost completely ignoring the latest know-how in weathering?" Especially as the particular influence of water on polymer ageing was proven and generally known before the latest issue of the standard was created. Standards are the products of experts. Furthermore, an important tool for shortening the artificial weathering has been carelessly given away.

For example of modern automotive coatings, it was shown in [22] that the test panels suffered faster loss of gloss outdoors in Miami, Florida, than during artificial weathering in a fluorescent UV lamp device employing uncontrolled relative humidity. In other words, in spite of available technical means, **none of the exposure cycles recommended in Part 3 of ISO 4892 is capable of meeting the requirements for simulating the natural weathering stress as either time-compressed or accelerated testing**.

The standard allows the use of higher relative humidity during the phases of irradiation only as an option. However, that means for a testing laboratory: "performing artificial weathering outside the standardised exposure cycles". This raises further questions: "Why was the standard developed in a manner that defies common sense, and why was one exposure cycle employing high relative humidity not agreed on and included in the standard?" All these questions can hardly be answered by persons not involved in creating the new version.

Obviously, the present market situation to the effect that the bulk of fluorescent UV lamp devices commercially available are poorly equipped to create such high relative humidity during irradiation was accorded greater consideration. Probably, suppliers and users were to be afforded the possibility of using weathering technology that is either antiquated or severely limited in its efficiency for artificial weathering tests that comply with international standards. As a result, testing laboratories see no reason to purchase or demand improved weathering technology.

Unlike the newly issued Part 2 (xenon lamps), Part 3 (fluorescent UV lamps) can only serve as a bad example and as a serious drag on the development of improved weathering technology. In this case, paint makers and coatings users, have been done a disservice. The results of weathering methods specified in Part 3 of ISO 4892 hardly lend themselves to correct predictions of the service lives of coatings.

9 References

[1] DIN EN ISO 877:1997-05
[2] VDA 75202: 2001-08
[3] DIN EN ISO 11997-1: 2006-04
[4] DIN 50018: 1997-06
[5] U. Schulz, Course „Coating technology“, Module 5, A 3, Vincentz Network, Hannover 2005, Seminar book
[6] DIN EN ISO 11997-2: 2006-04
[7] DIN 55665: 2007-01
[8] DIN EN ISO 2810: 2004-10
[9] VDI 3958 part 01: 1995-12
[10] U. Schulz, „Natürliches und künstliches Bewittern polymerer Werkstoffe“, TA Wuppertal, 05./06. Apr. 2006, Seminar book, Chapter 5
[11] DIN 50035-2, 1989-04
[12] G. Wypych, Handbook of Material Weathering 3rd Edition, ChemTec Publishing, Toronto 2003
[13] J.G. Calvert, J.N. Pitts, Photochemistry, John Wiley & Sons, New York 1967
[14] B. Ranby, F.J. Rabek, “Photodegradation, Photooxydation and Photostabilisation of Polymers”, John Wiley & Sons, London 1975
[15] P. Vink, “Degradation and Stabilisation of Polyolefins”, Appl. sci. Publ., London 1983
[16] A. Valet, “Lichtschutzmittel für Lacke”, Ed. U. Zorll, Vincentz, Hannover 1996
[17] M. Zäh, “Natürliches und künstliches Bewittern polymerer Werkstoffe”, TA Wuppertal 05./06.Apr. 2006, Seminar book, Chapter 10
[18] U. Schulz, G. Tjandraatmadja, Materialprüfung 45 (2003) 11/12, p.535-539
[19] J. Winkler, “Titandioxid”, Ed. U. Zorll, Vincentz, Hannover 2003
[20] G. Kämpf, G, Völz, A. Klaeren, XIV. FATIPEC-Congress, Congress book, 1978, p. 291-296
[21] U. Schulz, P. Trubiroha, “New Technologies of Industrial Finishing”, Congress book, Birmingham, 1998
[22] W. Rauth, S. Nowak, “35. Jahrestagung der GUS”, Pfinztal 2006, Congress book, p. 177-204
[23] A. Jurgetz, H. Rothbächer, a.o., Farbe & Lack 91 (1985) 10, p. 921-927
[24] W. Papenroth, P. Koxhold, Defazet 32 (1978) 3, p. 102-108
[25] H. Peters, XX. FATIPEC-Congress, Nizza 1990, Congress book 1978, p.292-296
[26] DIN 5031-1: 1982-03
[27] CIE-Publikation Nr. CIE85, 1. Edition, 1989, Technical Report „Solar Spectral Irradiance“
[28] M. McGreer, “Weathering Testing Guidebook”, Ed. Atlas Electric Devices Company, Pub.-No. 2062/098/200/AA/03/01
[29] H. K. Hardcastle III, ATLAS Library, www.atlas-mts.com, Apr. 2006
[30] Atlas Weathering Service Group, "Natural Weathering Testing" www.atlas-mts.com, Mar. 2008
[31] K. Dehne, “Natürliches und künstliches Bewittern polymerer Werkstoffe”, TA Wuppertal 01/02 Apr. 2004, Seminar book, Chapter 8
[32] A. Geburtig, “Natürliches und künstliches Bewittern polymerer Werkstoffe”, TA Wuppertal 05./06. Apr. 2006, Seminar book, Chapter 4
[33] ISO 9370: 1997-10
[34] W. Schweigart, “28. Jahrestagung der GUS”, Pfinztal 1998, Congress book, p. 200-204
[35] L.F.E. Jacques, Prog. Polym. Sci. 25 (2000) 1337-1362
[36] T. Nguyen, J. Martin, J.W. Byrd, Polymer Preprints 42 (2001) 1
[37] E. Klinke, D. Eisenbach, Farbe & Lack, 106 (2000) 3, p.112-124
[38] J. Lehmann, “Natürliches und künstliches Bewittern polymerer Werkstoffe”, TA Wuppertal 05/06 Apr. 2006, Seminar book, Chapter 6
[39] P. Trubiroha, V. Wachtendorf, A. Geburtig, “35. Jahrestagung der GUS” Pfinztal 2006, Congress book
[40] J.Boxhammer, D. Kokott, P. Trubiroha, Materialprüfung 35 (1993) 5, p. 143-147
[41] A. Schönlein, “33. Jahrestagung der GUS”, Pfinztal 2004, Congress book, p 121-135
[42] P. Trubiroha “Natürliches und künstliches Bewittern polymerer Werkstoffe”, TA Wuppertal 05/06 Apr. 2006, Seminar book, Chapter 1
[43] J. Sampers, Polymer Degradation and Stability 76 (2002) 544-465
[44] T. Helmen, Y. Fakhoury, Farbe & Lack 99 (1993) 5, p. 431-440

Ulrich Schulz: Accelerated Testing:
Nature and Artificial Weathering in the Coatings Industry

ISBN: 978-3-86630-908-1

[45] W. Marquardt a.o., Tellus 53B (2001) p. 529-545
[46] G. T. Wolff, W. R. Rodgers a.o., J. Air Waste Manage. Assoc. 40 (1990) 12, p. 1638
[47] T. Reichert, "Natürliches und künstliches Bewittern polymerer Werkstoffe", TA Wuppertal 05/06 Apr. 2006, Seminar book, Chapter 8
[48] U. Schulz, "Natürliches und künstliches Bewittern polymerer Werkstoffe", TA Wuppertal 05/06 Apr. 2006, Seminar book, Chapter 7
[49] U. Schulz, V. Wachtendorf, A. Geburtig, ECJ 05 (2004) p. 26-30
[50] Allunga Exposure Laboratory, www.allunga.com (Apr. 2007)
[51] U. Schulz, K. Witt, Farbe & Lack 103 (1997) 8, p. 51-63
[52] D. Kokott, "Natürliches und künstliches Bewittern polymerer Werkstoffe", TA Wuppertal 01/02 Apr. 2004, Seminar book, Chapter 10
[53] DIN EN ISO/IEC 17025:2005-08
[54] Anonym, ATLAS Library, Chapter 10, www.atlas-mts.com, Apr. 2007
[55] SAE J1976: 2002-03
[56] U. Schulz, A. Geburtig, M. Crewdson, J. Stephenson, Galvanotechnik 1/2006, p. 166-172
[57] T. Reichert, 1st European Weathering Symposium, Prague, Sept. 2003, CEEES Publication No 5, 2004, p. 341-355
[58] NOAA Climate Portal, www.noaa.com/wx.html (Oct. 2006)
[59] A. Lüthi, D. Fäh, W. Wachtendorf, A. Geburtig, 1st European Weathering Symposium, Prague, Sept. 2003, CEEES Publication No 5, 2004, p. 81-93
[60] V. Wachtendorf, U. Schulz, P. Trubiroha, 1st European Weathering Symposium, Prague, Sept. 2003, CEEES Publication No 5, 2004, p. 301-318
[61] DIN EN ISO 1514: 2005-02
[62] Q-Lab, Product Information, 2004
[63] J. Boxhammer, XXII. Colloquium of Danubian Countries on Natural and Artificial Ageing of Polymers, Berlin 2001, Proceedings, V 2
[64] J. Chin, T. Nguyen, E. Byrd, J. Martin, 2nd European Weathering Symposium, Gothenburg, June 2005, CEEES Publication No 6, 2005, p. 105-130
[65] VDA working group "Kurzbewitterung", private information till 2004
[66] Patent DE 10 2004 048 028 B4, 28.09.2004/24.08.2006
[67] J. L. Gerlock u.a., JCT, Vol.75, No. 936, Jan. 2003
[68] A. Riedl, SunSpots, "Material Testing Product and Technology News", Vol. 35, Issue 75
[69] B. Klemann, SunSpots, "Material Testing Product and Technology News", Vol. 35, Issue 75
[70] E. Pohle, L. Kranner, 1st European Weathering Symposium, Prague, Sept. 2003, CEEES Publication No 5, 2004, p. 17-36
[71] A. Schönlein, "36. Jahrestagung der GUS", Pfinztal 2007, Congress book, p. 327-339
[72] A. Geburtig, private Information, Mar. 2007
[73] K. Horn, "Bewitterung von Kunststoffen im Fahrzeugbau", SKZ Würzburg 2001, Seminar book, Chapter F
[74] ATLAS, "Network of Weathering, Product and Services", CD 2002
[75] Weiss-Umwelttechnik GmbH, Product information "Global-UV-Test Unit Model UV200", 2000
[76] G. R. Pilcher, G. N. Van de Streek a. o., 1st Symposium of Life Prediction, 1997, Breckenridge/CO
[77] A. Schönlein, "Bewitterung von Kunststoffen im Fahrzeugbau", SKZ Würzburg 2005, Seminar book, Chapter J
[78] A. Geburtig, "Natürliches und künstliches Bewittern polymerer Werkstoffe", TA Wuppertal 05/06 Apr. 2006, Seminar book, Chapter 13
[79] A. Geburtig, V. Wachtendorf u.a. 2nd European Weathering Symposium, Gothenburg, June 2005, CEEES Publication No 6, 2005, p. 145-154
[80] J. Boxhammer, 2nd European Weathering Symposium, Gothenburg, June 2005, CEEES Publication No 6, 2005, p. 133-144
[81] SAE J1960:2004-10
[82] T. Raabe, XXII. Colloquium of Danubian Countries on Natural and Artificial Ageing of Polymers, Berlin 2001, Proceedings, V. 6
[83] P. Trubiroha, A. Geburtig, V. Wachtendorf, "Bewitterung von Kunststoffen und Lacken in der Automobilindustrie", SKZ Würzburg, 2007, Seminar book, part A
[84] U. Schulz, V. Wachtendorf, "Aufklärung der Ursachen für Lackschäden", Study, part II "Saure Beaufschlagungen", Düsseldorf, Sept. 2002
[85] U. Schulz, Course "Coating technology", Module 5, A 3, Vincentz-Network, Hannover 2005, Seminar book
[86] VDI 3958, Bl. 12: 2004-02
[87] M. Palm, B. Carlsson, JCT 1 (Jan. 2002), Vol. 74, No. 924, S. 69-74
[88] J. Boisseau, D. Cambell u.a., 1st European Weathering Symposium, Prague, Sept. 2003, CEEES Publication No 5, 2004, p.137-148
[89] W. Rauth, Meeting protocol of the VDA working group "Kurzbewitterung", July 2004
[90] U. Schulz, P. Trubiroha, U. Schernau a. o., Progr. in Org. Coatings, 40 (2000), S. 151-156

[91] Q-Panel Lab Products, www.q-panel.com
[92] DIN EN ISO 4628-1: 2004-01
[93] ISO 105-A02:1993-09
[94] Byk-Gardner Product catalogue, Geretsried 2005
[95] http//de.wikipedia.org/wik/rangkorrelationskoeffizient
[96] A. Geburtig, V. Wachtendorf, P. Trubiroha u.a., "36. Jahrestagung der GUS", Pfinztal 2007, Congress book, S. 247-256
[97] E. Klinke, Course "Coating technology", Module 5, A 6, Vincentz-Network, Hannover 2005, Seminar book
[98] DIN ISO EN 2409: 1994-10
[99] A. Geburtig, U. Schulz, M. Mechtel, T. Klimmasch, 7th Nürnberg Congress, 7-8 Apr.2003, Congress book
[100] E DIN 55668: 2003
[101] DIN EN ISO 4892-2:2006-05
[102] DIN EN ISO 4892-3:2006-05
[103] W. Rauth, S. Nowak, 3rd European Weathering Symposium, Krakow, Sept. 2007, CEEES Publication No 8, 2007, p. 275-282
[104] J. Pietschmann, M. Crewdson, G. Pausch, 3rd European Weathering Symposium, Krakow, Sept. 2007, CEEES Publication No 8, 2007, p. 233-242

Index

A

B

C

Ulrich Schulz: Accelerated Testing:
Nature and Artificial Weathering in the Coatings Industry

ISBN: 978-3-86630-908-1

D

E

F

G

H

I

J

K

L

M

N

O

P

Q

R

S

X

Y

www.ingramcontent.com/pod-product-compliance
Lightning Source LLC
LaVergne TN
LVHW081330021026
849161LV00008BA/1835